Macromolecules Regulating Growth
and
Development

Macromolecules Regulating Growth and Development

The Thirtieth Symposium
The Society for Developmental Biology

Seattle, Washington, June 17–19, 1971

Macromolecules Regulating Growth and Development

Edited by

Elizabeth D. Hay

Department of Anatomy
Harvard Medical School
Boston, Massachusetts

Thomas J. King

Division of Cancer Research Resources and Centers
National Cancer Institute
Bethesda, Maryland

John Papaconstantinou

Biology Division
Oak Ridge National Laboratory
Oak Ridge, Tennessee

1974

ACADEMIC PRESS *New York and London*

A Subsidiary of Harcourt Brace Jovanovich, Publishers

ACADEMIC PRESS, INC.
111 Fifth Avenue, New York, New York 10003

United Kingdom Edition published by
ACADEMIC PRESS, INC. (LONDON) LTD.
24/28 Oval Road, London NW1

LIBRARY OF CONGRESS CATALOG CARD NUMBER: 55-10678

ISBN 0–12–612973–8

PRINTED IN THE UNITED STATES OF AMERICA

Contents

I. Regulatory Factors in the Selective Growth of Mammalian Cells

Role of the Cell Surface in Growth and Transformation

MAX M. BURGER

Epidermal Growth Factor:
Chemical and Biological Characterization

STANLEY COHEN AND JOHN M. TAYLOR

Tumor Angiogenesis: Role in Regulation of Tumor Growth

JUDAH FOLKMAN

II. Formation and Organization of Plant Cell Walls and the Plasma Membrane

III.   Organization and Expression of Genetic Information

Organization of DNA and Proteins in Mammalian Chromosomes

ELTON STUBBLEFIELD

The Multiple Relations of tRNA to Metabolic Control

ROBERT M. BOCK

Total Synthesis of Transfer RNA Genes
H. G. KHORANA

Biosynthesis of Bacterial Ribosomes

MASAYASU NOMURA

The Structural Basis of Selective Gene Expression
in Eukaryotes

BRIAN J. McCARTHY AND MICHEL JANOWSKI

Gene Manipulation: Progress and Prospects

ETHAN SIGNER

Macromolecules Regulating Growth and Development

The 30th Symposium was held at Seattle, Washington, June 17–19, 1971. The Society gratefully acknowledges the efficiency of the host committee, the hospitality of The University of Washington and the support from the National Science Foundation.

Contributors and Presiding Chairmen

Chairman. Arthur H. Whiteley, Local Committee on Arrangements. Welcome

Convener of Symposium. Folke Skoog, University of Wisconsin, Madison, Wisconsin. Introductory Remarks

I. Synthesis and Organization of Chromosomal Components

Chairman. Benjamin D. Hall, University of Washington, Seattle, Washington

T. ELTON STUBBLEFIELD, M. D. Anderson Hospital, University of Texas, Houston, Texas

H. GOBIND KHORANA, Massachusetts Institute of Technology, Cambridge, Massachusetts

ETHAN SIGNER, Massachusetts Institute of Technology, Cambridge, Massachusetts

II. Formation and Organization of Plant Cell Walls

Chairmen. Robert Cleland, University of Washington, Seattle, Washington and Anton Lang, Michigan State University, East Lansing, Michigan

D. JAMES MORRÉ, Purdue University, Lafayette, Indiana

DEREK T. A. LAMPORT, Michigan State University, East Lansing, Michigan

MAARTEN CHRISPEELS, University of California at San Diego, La Jolla, California

III. Characterization of Anabolic Organelles

Chairman. Brian J. McCarthy, University of Washington, Seattle, Washington

ANDREW A. BENSON, University of California at San Diego, La Jolla, California

ROBERT M. BOCK, University of Wisconsin, Madison, Wisconsin

MASAYASU NOMURA, University of Wisconsin, Madison, Wisconsin

Discussions. On selected topics prearranged by participants with STEPHEN HAUSCHKA

IV.   Regulatory Factors in the Selective Growth of Mammalian Cells

Organizer and Chairman.   Gerald C. Mueller, University of Wisconsin, Madison, Wisconsin

BRIAN J. McCARTHY, University of Washington, Seattle, Washington

MAX M. BURGER, Princeton University, Princeton, New Jersey

STANLEY COHEN, Vanderbilt University, Nashville, Tennessee

JUDAH FOLKMAN, Boston City Hospital, Boston, Massachusetts

BERT W. O'MALLEY, Vanderbilt University School of Medicine, Nashville, Tennessee

Preface

The 30th Symposium of the Society for Developmental Biology was organized by Folke Skoog with the help of Robert Cleland, Benjamin D. Hall, Anton Lang, Brian J. McCarthy, and Gerald C. Mueller, and was held at the University of Washington in Seattle, June 17–19, 1971. Arthur H. Whiteley was Chairman of the Local Committee on Arrangements. As has been the custom in the past, all of the manuscripts submitted by the speakers were accepted for publication in this volume. The manuscript by Ethan Signer deals with social issues which are difficult to evaluate and reflects his personal opinion, not necessarily that of the Editors, the Society, or the Publisher. In the Appendix to the 31st Symposium of the Society the reader will find discussions presenting other viewpoints of the social implications of genetic engineering.

The Society would like to express its appreciation to the University of Washington and the host committee for their hospitality, and to the National Science Foundation for support of this meeting, which was attended by more than 300 Society members, students, and guests.

ELIZABETH D. HAY
THOMAS J. KING
JOHN PAPACONSTANTINOU

I. Regulatory Factors in the Selective Growth of Mammalian Cells

Role of the Cell Surface in Growth and Transformation

MAX M. BURGER

Department of Biochemical Sciences, Princeton University,
Princeton, New Jersey

I. INTRODUCTION

Why should the surface membrane have any role or function in the regulation of either cell growth or development? Otto Nägeli (Nägeli and Cramer, 1855) created the concept of a plasma membrane about 120 years ago and thought of it primarily as a permeability barrier. For the greater part of this period it was considered to be simply a bag which kept vital internal components within the cell and protected them from damaging influences from without. The membrane guaranteed thereby an optimal internal microenvironment for each cell. During early stages of evolution this may indeed have been the principal function of the surface membrane. Later on, however, the surface membrane developed a series of specialized functions. It became involved in locomotion, provided hormone receptors, antigenic recognition sites, excitability, etc., i.e., functions that are involved in communication and recognition.

If the cell surface appears to have the function of the sensory organelle for a single cell, it seems that it may have used such a function also to its advantage in regulating cell division and cell development according to the needs dictated by the cellular environment.

3

There are situations in which cells want to divide actively (e.g., wounds, during embryonic growth, after immunologic stimuli) as well as situations in which such proliferation is no longer required (e.g., in most adult or differentiated cells that are carrying out a specific function). It would therefore be to the cell's advantage if the plasma membrane had a direct role in growth control since it is the surface that will first detect changes in cellular environment whether these changes are fluctuations in the concentration of soluble communication agents, or the degree of proximity or interaction with neighboring cells.

Besides such a role of the cell membrane in controlling growth based on environmental influences surrounding the cell, we postulate as a working hypothesis an interrelation between the surface membrane and the interior of the cell, e.g., the nucleus. Within many tissues, the nuclear DNA per cell and the number of nuclei are fairly stable. In order to keep such ratios within relatively small limits of deviation, a feedback control between some events during the cell cycle monitoring cell division and nuclear division may therefore be postulated. After division and segregation of the nuclear material, the cell itself should divide before going into the next cycle of replication of genetic material. This is presumably achieved by the centrioles, which may set the division furrow. Only after division has occurred should the cell be permitted to enter the next cycle, which leads, among other things, to the duplication of genetic material. Basically any process prior to or during mitosis could be instrumental in such a control loop, but the surface membrane seems to be the best candidate since it is the actual separation of the two newly formed nuclei into two distinct cytoplasms which prevents the formation of polynucleated cells. This separation is achieved by the surface membrane, and therefore we postulate as a working hypothesis that the plasma membrane is the carrier of the information for the nucleus to enter in the new round of the next cell cycle, giving rise to another chromosomal set, and this in turn would trigger thereafter a new cell division (see Fig. 1). In its basic outline, this concept is similar to the one suggested by Jacob *et al.* (1963) for bacteria, keeping in mind that the genetic material in animal cells seems to be separated from the environment by the plasma and the nuclear membrane but in bacteria only by the plasma membrane.

II. LOSS OF DENSITY-DEPENDENT GROWTH CONTROL IN TRANSFORMED CELLS AND CELL SURFACE ALTERATIONS

Many transformed cell cultures grow to higher cell densities, at identical serum concentrations and pH, than do their untransformed parental

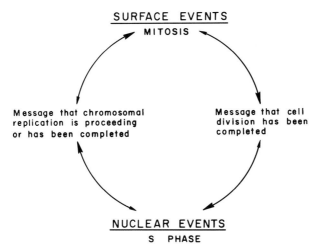

FIG. 1. Working hypothesis for a possible cyclic communication between the cell membrane and the cell nucleus. From Fox *et al.* (1971).

counterparts. This pertains to virally, chemically, spontaneously, and X-ray-induced transformed cells and applies to not only fibroblasts, but also to epithelial cells (Borek *et al.*, 1973). Since for some pairs of transformed and untransformed cells a correlation between the saturation density *in vitro* and tumorigenicity *in vivo* was found, the escape from density-dependent inhibition of growth (contact inhibition of growth) in cell cultures was considered to be a significant *in vitro* criterion for growth aberration *in vivo* (Aaronson and Todaro, 1968). Exceptions to this correlation have been reported (Sanford *et al.*, 1967) and will be further found. Furthermore, a serious objection will always be the question whether such artificial *in vitro* parameters have any bearing on the *in vivo* growth behavior, but we have to work with this rather crude, and admittedly far from ideal, growth assessment until better quantitation of the growth aberration seen *in vivo* becomes available.

A. Surface Alteration Detected after Transformation of Cell Cultures

1. Relation between Contact Inhibition of Growth and Surface Alteration. Ambrose and his co-workers originally detected a growth inhibitory effect by a lipase preparation from wheat germ (1961). Aub and his group established that this preparation agglutinated some tumor cells but in general no untransformed cells (Aub *et al.*, 1963, 1965a,b). We then purified the agglutinating impurity in the wheat germ lipase and

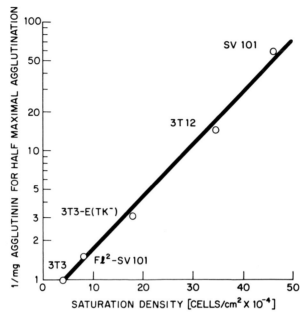

FIG. 2. Correlation between loss of density-dependent inhibition of growth and agglutinability. These results were given earlier in a table by Pollack and Burger (1969).

found that it agglutinated all tissue culture lines transformed with DNA tumor viruses as well as chemically transformed cells (Burger and Goldberg, 1967).

Correlation between the degree of escape from contact inhibition of growth and agglutinability was further support for a possible significance of this surface alteration (Burger, 1970a). Pollack (Pollack *et al.*, 1968; Pollack and Burger, 1969) isolated a line (Fl²SV101) from virally transformed cells (SV101) which was selected for a reversion to normal growth properties; i.e., these cells displayed again contact inhibition of growth. Together with the reversion in growth control properties, these cells were also reverting from their surface alteration and turned out to agglutinate essentially as poorly as untransformed tissue culture cells (Fig. 2).

The mirror image experiment was also carried out. Cells were selected among the agglutinating transformed cell lines which would not agglutinate, i.e., have an unaltered normal cell surface, and then the question was asked how these cells grew, i.e., whether they would also have reverted to normal growth control. Such clones were indeed found, although the reversion of growth control was not complete (Ozanne and Sambrook, 1971).

TABLE 1

AGGLUTINABILITIES OF 3T3 MOUSE FIBROBLASTS INFECTED WITH
HOST RANGE MUTANTS OF POLYOMA VIRUS[a]

| Cells | Agglutinin necessary for half-maximal agglutination | |
	Wheat germ agglutinin (μg/ml)	Concanavalin A (μg/ml)
3T3[b]	300–500	600
3T3 polyoma virus transformed	40	120
3T3 infected by Py-virus mutant NG-18	200–300	600
NG-23	200–300	—
NG-59	200–300	—
HA-33	200–300	—
3T3 infected by Py-virus wild type	20–40	150

[a] Taken in part from Benjamin and Burger (1970).
[b] Well contact-inhibited 3T3 cells sometimes require up to 800 μg/ml.

2. *Studies with Mutants of Transforming Viruses.* Mutants of polyoma virus that were unable to infect and transform normal 3T3 fibroblasts but grew on 3T3 cells transformed with wild-type virus (Benjamin, 1970) were tested. Such host-range mutants were unable to bring about the same surface change usually seen after infection with the wild-type virus (Benjamin and Burger, 1970) even though other functions indicating that infection took place were still expressed (Table 1).

A temperature-sensitive mutant of polyoma virus which infected and transformed the host cell only at its permissive temperature (32°C) was unable to bring about the usual alteration (agglutination) seen after infection with the wild-type virus if infection was carried out at the nonpermissive temperature (39°C). If a cell line, earlier transformed at 32°C with the mutant virus, was shifted to the higher temperature (39°C) nonpermissive for infection and successful transformation, it lost its surface alteration for as long as it was kept at the higher temperature. After a shift to the permissive temperature (32°C) again, it reverted to the agglutinable or transformed cell surface configuration (Fig. 3), indicating that the virus, which made a temperature-sensitive macromolecule, has to be present at all times to keep the cell surface in its agglutinable state (Eckhart *et al.*, 1971).

3. *Possible Significance in Vivo.* From hapten inhibition studies, we concluded that the disaccharide of N-acetylglucosamine, i.e., di-N-acetylchitobiose, was part of the receptor molecule for the agglutinin we isolated from wheat germ lipase. Shier (1971) immunized mice with di-N-

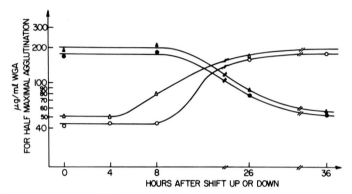

FIG. 3. Appearance and disappearance of the surface alteration after a shift in temperature in ts-3-transformed BHK cells. ▲——▲, Clone 7-C shifted from 39° to 32°C; ●——●, clone 1 shifted from 39° to 32°C. △——△, clone 1 shifted from 32° to 39°C; ○——○, clone 7-C shifted from 32° to 39°C. From Eckhart *et al.* (1971).

acetylchitobiose covalently attached to polyaspartate and found that they were protected to some degree against tumor if myeloma cells were injected or against formation of tumor if challenged with a chemical carcinogen. This result, although directly deducible from the *in vitro* results, is still impressive since erythrocytes and granulocytes agglutinate quite well (Aub *et al.*, 1965a) and one might have expected some serious side reactions in such an experiment as autoimmune reactions or inhibition of fast division in regenerating tissues (Burger, 1970a).

B. Possible Mechanisms Leading to the Surface Alteration

1. Observations. We originally considered three possibilities that could lead to this alteration: (a) *De novo* formation of the alteration leading to the agglutinable state; the information for the formation of the site would be directly coded by the virus; (b) activation of the formation of the host-coded receptor sites or alterations of the host-coded receptor sites by the transforming virus; (c) a rearrangement in the architecture of the cell surface leading to the agglutinable state whereby the receptor sites for the agglutinins would already be present in the normal cell surface to the full extent. Whether this is achieved by a simple exposure of the receptor site for the agglutinin or by some more complicated mechanism remains open.

The first possibility could be ruled out for various reasons, not the least being the fact that many cells that have not been transformed by a virus are agglutinable. Such "normal" tissue culture cells, e.g., baby hamster kidney (BHK), which agglutinate to a minor extent have, as predicted, lost some of their growth control. Furthermore, spontaneously, chemically, and X-ray transformed cells have been found to display this surface transition, i.e., cells that were presumably not triggered into transformation by a virus. Finally, it would be unlikely that some of the small oncogenic viruses, carrying 5–9 genes only, would have enough information to code for the multitude of agglutinin receptor sites which are presently discovered in transformed cells.

In order to distinguish between the two last possibilities, one has to isolate the receptor site and assess it quantitatively in the parent and transformed cell. We have partially isolated such a fraction from transformed cell surfaces (Burger, 1970a), compared yields from transformed and parent cells and found that approximately the same amount was present in the polyoma virus-transformed baby hamster kidney (Py-BHK) and BHK cells. Dr. V. K. Jansons has carried further the isolation of the wheat germ agglutinin (WGA) site fraction (Burger and Jansons, 1972) finding about four glycoproteins on SDS disc gel electrophoresis with molecular weights in the neighborhood of 25,000–45,000. This fraction seems to be either free of any glycolipids or contains only traces of lipids. The material isolated so far (from chemically induced L1210 mouse leukemia cells) is specific for WGA only and seems not to interact with concanavalin A, another lectin with a tendency to agglutinate transformed cells preferentially. Complement-fixing antibodies to this WGA receptor fraction were directed not toward the carbohydrate determinant of the WGA but toward another part of the receptor fraction. They lysed only the lymphocytic leukemia cells, not normal lymphocytes. These antibodies prevent also the agglutination of virally transformed Py-BHK and Py-3T3 cells by WGA, presumably by specifically covering the WGA receptor sites. Agglutination by concanavalin A was not inhibited or inhibited much more poorly.

In view of the fact that not only transformed, but also untransformed, cells seemed to have receptor sites for WGA, we reasoned that enzymatic treatment of the surface may make these sites available by whatever mechanism is theoretically conceivable, the most simple being uncovering of the sites which would be cryptic in normal cells. Very mild treatment of all sorts of untransformed cells with proteolytic enzymes turned out to bring about agglutinability in untransformed cells which was quantitatively and qualitatively essentially identical with that found in their virally transformed counterparts (Burger, 1969). Evidence is accumulating that this surface alteration brought about by proteolytic enzymes is more general and applies to other surface antigens as well (Hakomori et al., 1967; Inbar and Sachs, 1969a; Sela et al., 1970; Häyry and Defendi, 1970). Some of these results are reflections of a higher binding of

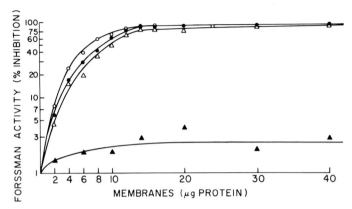

FIG. 4. The ordinate reflects the degree of Forssman antigens expressed on isolated hamster cell surface membranes. Increasing amounts of isolated membranes
(abscissa) bound increasing amounts of Forssman antibodies in a standard inhibition
test. ▲——▲, Baby hamster kidney (BHK) cell membranes; △——△, BHK cell
membranes treated with 0.008% chymotrypsin for 6 minutes; ●——●, polyoma
virus-transformed BHK cell membranes; ○——○, polyoma virus-transformed BHK
cell membranes treated with the same amount of chymotrypsin. For details, see
Burger (1971a).

antibodies to the transformed and the trypsinized normal cell surface
and can therefore not be dismissed as differences not occurring at the
binding site for the antibody. This applies primarily to the binding of
fluorescent antibodies (Häyry and Defendi, 1970) and to some degree
also to the complement lysis assay (Hakomori *et al.*, 1967). We have
recently added another antigen to this list which occurs in virally transformed cells and can be made available for antibody binding in the untransformed tissue culture cell after mild proteolytic treatment (Burger,
1971a) (Fig. 4).

2. *The Open Questions.* First: If the receptor fraction is ultimately
purified and is still comparable in the amounts found in transformed and
untransformed cells, the question remains whether they are identical molecules. Even though they may have identical physical chemical properties
and are therefore isolated in the same fractions, they may slightly differ
in their protein backbone or in the distribution and number of carbohydrate side chains or even in the carbohydrate side chains themselves although a small sequence within these side chains containing the receptor
site could be identical.

Second: We still do not know how to explain the effect of the mild
proteolytic treatment on the agglutinability of the untransformed cells
in molecular terms. The proteolytic activity has been shown to be a re-

quirement, since inhibition of proteolysis or inactivated proteases do not act. Furthermore, the activity is not due to some phospholipase or to other impurities since specific inhibition of trypsin, e.g., by the soybean trypsin inhibitor tosyl phenylalanine chloromethyl ketone (TPCK) or by ovomucoid, inhibits this effect. There is no specificity for any given peptide linkage since so far any protease tested will bring about agglutinability of untransformed fibroblasts. For a long time we have attempted to isolate a protein that would have come off the surface under the mildest conditions just leading to a detectable agglutinability (for properly washed cells this can be as low as 1–2 μg of trypsin per milliliter for 30 seconds) and since various approaches have not yielded any detectable protein or peptide release, we thought of the proteolytic effect as a cleavage of a peptide linkage, and thereby a rearrangement in the surface membrane that would lead to the agglutinable state. Sheinin's group (Onodera and Sheinin, 1970; Sheinin, 1972) working with longer periods of protease treatment and higher concentrations of protease has demonstrated interesting differences between transformed and untransformed tissue culture cells. Drs. H. P. Schnebli and J. R. Sheppard have recently found, however, that heavily labeled cells will release some proteinaceous material even under mild proteolytic conditions. Whether this release is directly instrumental in bringing about agglutinability or whether it is just an insignificant by-product of the rearrangement induced by this small amount of protease remains one of the central questions.

C. Preliminary Indications for a Functional Implication of This Surface Alteration

1. Proteolytic Enzymes Lead to a Temporary Escape from Growth Control. If mild proteolytic treatment of untransformed cells can give rise to a surface alteration similar to that seen in transformed cells, growth control of such cells might also become similar to that in transformed cells, as long as the surface alteration persists. This prediction rests on the assumption that the cell surface is mediating growth control, an assumption for which no relevant evidence can be cited so far. A brief proteolytic treatment of untransformed mouse fibroblasts can indeed bring about escape from density-dependent inhibition of growth (Burger, 1970b). Since such cells return to their nonagglutinable state 6 hours after this treatment, one would expect them to return to normal growth control after that time, and they indeed will go only through at most one round of cell division. A second proteolytic treatment thereafter has been shown by K. D. Noonan to give another growth response of similar magnitude (Fig. 5).

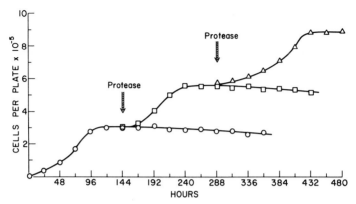

FIG. 5. Overgrowth of 3T3 mouse fibroblasts after a 5-minute treatment with 5 μg of Pronase per milliliter. Unpublished results of K. D. Noonan; for details, see Burger (1970b).

The proteolytic effect seems to be restricted to the cell surface since a brief incubation with proteolytic enzymes attached to beads and not phagocytosed (controlled with tritium-labeled beads) could also bring about a temporary release from growth control as well as agglutinability (Fig. 6). This experiment contributes an answer to another question: Does the whole surface have to be treated with a proteolytic enzyme or is it sufficient that some parts only are reached by the protease? Since not enough beads were used to cover the entire upper surface; since the culture was not shaken around during incubation and the beads therefore were not rolling around on the cell surface, since the beads could not reach the cell side that was facing the substratum and since control experiments showed that essentially no protease was released from the

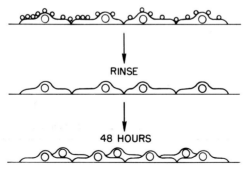

FIG. 6. Overgrowth after treatment of 3T3 mouse fibroblasts for 5 minutes with Pronase or trypsin covalently linked to beads (M. M. Burger, unpublished observation).

beads into the medium which could have reached other parts of the cell surface, we think that the proteolytic action at only a few locations on the cell surface was able to bring about escape from growth control. Sefton and Rubin (1970) have shown independently that chick embryo fibroblasts also show overgrowth to a larger extent if they are subjected to a mild trypsin treatment, and initiation of a new cell cycle has also been demonstrated by Vasiliev et al. (1970) using degradative enzymes acting on the cell surface. Such cells treated with trypsin also become better agglutinable with WGA (Burger, 1970a).

If proteolytic enzymes can simultaneously bring about the surface alteration usually seen in transformed cells and escape from contact inhibition of growth, it may be concluded that some transformed cells may have increased proteolytic enzymes that act on membrane precursors inside the cells, or on their surface, or are secreted into the medium. Preliminary results from experiments done four years ago supported the notion that—at least in L1210—cells seem to contain surface proteases that can bring about the release of untransformed 3T3 fibroblasts from contact inhibition of growth if in close contact (Burger, 1971b).

During a study on interactions between transformed and normal cells in tissue culture, stimulating effects of transformed cells upon confluent

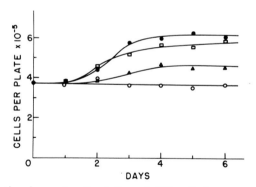

FIG. 7. Proportional growth stimulation of 3T3 cells by exposure to increasing numbers of L1210 leukemia cells. Various numbers of L1210 leukemia cells were added to 3T3 mouse fibroblasts for 13 hours in conditioned medium. L1210 cells were removed by pouring them off and rinsing with dilute EDTA solutions in Ca^{2+} and Mg^{2+}-free phosphate-buffered saline. Controls (O——O) were treated identically. The controls received the same conditioned medium, which contained 3% calf serum and had been depleted for 3 days on a 3T3 culture. Saturation densities were determined by counting cells released with 0.01% trypsin in a Levy-Hausser chamber. At all three concentrations of L1210 leukemia cells added (△——△, 10^4 cells; □——□, 10^5 cells; ●——●, 10^6 cells, less than one cell in a hundred 3T3 cells remained on the plates after rinsing. From Burger (1971b).

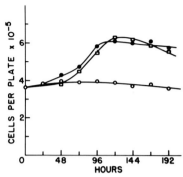

Fig. 8. Dependence of growth stimulation by leukemic cells on time of exposure. The same conditions were used as described in Fig. 7; 10^6 L1210 cells were added. Incubations shorter than 3 hours have not yet been done. The amount of proteolytic activity secreted during 3 hours is below the minimal concentration required for growth stimulation, as determined with the most active commercially available protease. Since we found that proteolytic inhibitors (e.g., tosyl arginine methyl ester and tosyl phenylalanine chloromethyl ketone) prevented the stimulatory effect of L1210 cells and since L1210 membrane preparations contain proteolytic activity and can stimulate growth of 3T3 cells, we propose that L1210 cells stimulate growth of 3T3 fibroblasts by means of surface-located proteases when the cells make contact. From Burger (1971b). □——□, Exposed to L1210 for 10 hours; ●——●, exposed to L1210 for 3 hours; ○——○, control, mock-treated with conditioned medium.

3T3 mouse fibroblasts were noticed. The best results were achieved when L1210 leukemia cells were used as the stimulating cells, since they could be layered over the 3T3 cells and removed fairly easily after a given time. They were also easily distinguished on morphological grounds (primarily size) from 3T3 fibroblasts. Figure 7 shows that the growth response was dependent on the number of leukemia cells added, and Fig. 8 shows that an exposure interval of 3 hours was sufficient to bring about the escape from density-dependent inhibition of growth. Culture medium from L1210 cells grown in tissue culture gave only a very poor response (10–20% higher saturation density) while ascitic fluid from mice inoculated intraperitoneally with L1210 cells resulted in almost a doubling of the saturation density (Fig. 9). In a further control, mouse lymphocytes (devoid of macrophages), the normal parent cells of L1210 leukemia cells, were incubated with the same overlaying assay, and no growth stimulation could be observed among the 3T3 fibroblasts.

Since tissue culture medium from L1210 cells contained only very little proteolytic activity while ascites fluid had 5- to 10-fold more, we thought that the growth stimulatory effect could be due to proteolytic enzymes. A recent more detailed study by Mr. Robert Remo indicates that the growth-stimulatory effect is not due, however, to proteolytic enzymes

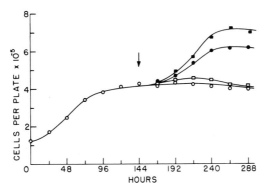

FIG. 9. Evaluation of the growth stimulatory effect of L1210 cells upon resting 3T3 mouse fibroblasts. ●——●, Peritoneal fluid, 1 ml, withdrawn together with L1210 cells from a 6-day ascites mouse carrying these cells, and conditioned 3T3 medium gave a growth-stimulatory effect. ○——○, Medium from 3-day-old L1210 leukemia cells grown in tissue culture did not give rise to the growth stimulus, □——□, L1210, 3 days old, in conditioned medium, seem not to secrete relevant amounts of proteases. The growth-stimulatory activity found for intact L1210 cells (see Figs. 7 and 8) seems to reside in isolated surface membranes from L1210 cells; ■——■, cells exposed to L1210 surface for 3 hours.

secreted into the medium by the transformed cell, but rather by prote-olytic enzymes located in the surface membrane and acting when the leukemic cell makes contact with the untransformed 3T3 fibroblast (see legend to Fig. 8). Inhibitors of proteolytic enzymes like tosyl arginine methyl ester (TAME) and tosyl phenylalanine chloromethyl ketone (TPCK) were able to block the growth stimulation seen after exposure of 3T3 fibroblasts to L1210 cells. The concentrations necessary for inhibition of growth induction by L1210 cells were below those necessary for inhibition of 3T3 growth prior to confluency, but by less than an order of magnitude. The studies are preliminary and will have to be confirmed. As seen in Fig. 9, growth stimulation did not require intact L1210 cells, since purified and washed plasma membranes of L1210 cells were able to stimulate even better (3-fold better than on a protein basis). To what degree these membranes were totally washed off again from the 3T3 cells is not yet exactly known. About 5–15% tritium-labeled membranes seemed to be irreversibly attached to 3T3 cells after 3 hours.

The growth stimulatory effect found with L1210 cells is not necessarily a general one, however, since virally transformed cells, sparsely seeded, seemed to be contact-inhibited by the confluent fibroblast layer. More detailed studies will be required before the significance of our preliminary studies upon other transformed/untransformed tissue culture systems or the situation *in vivo* can be evaluated.

2. Covering of Transformed Cell Surfaces with Nonagglutinating Lectin Brings about a Return to Normal Growth Control. If a brief proteolytic treatment of cells can lead to agglutinability and escape from growth control in normal cells, the question arises whether by bringing about a nonagglutinable state in transformed cells, one could enable transformed cells to regain their normal growth control. An experiment of this type was carried out already when shifting BHK cells transformed by a temperature-sensitive polyoma virus mutant from its permissive temperature (32°C) to its nonpermissive temperature (39°C) whereby both the agglutinability disappeared and the cells began to grow somewhat like untransformed cells (swirls) (Eckhart *et al.*, 1971). Furthermore, Pollack's variant cell (Fl²SV101), which was selected for density-dependent inhibition of growth from a SV40-transformed cell line (Pollack *et al.*, 1968), turned out to have gained back the nonagglutinable cell surface typical for untransformed cells (Pollack and Burger, 1969). A similar variant which has lost its agglutinability for concanavalin A has been described by Sachs' group (Inbar *et al.*, 1969). In order to achieve a nonagglutinable state in transformed cells, one may simply cover up the surface receptors with what we assume to be monovalent nonagglutinating pieces of an agglutinin. Gepner and Steinberg (1973) have independently found conditions that will produce a nonagglutinable preparation of concanavalin A by treating it with trypsin, and we had some success with chymotrypsin. Such preparations not only fail to agglutinate transformed cells, but also inhibit agglutination by native (presumably bivalent) agglutinin if the cells are preincubated with this presumably monovalent agglutinin.

Addition of such a trypsinized agglutinin preparation can bring about density dependent inhibition of growth in transformed cells for as long as the carbohydrate side-chain sites specific for the agglutinin are covered (Burger and Noonan, 1970). If they are uncovered with the help of the specific carbohydrate hapten, the transformed cells resume their normal growth pattern—i.e., they continue to grow to high saturation densities (Fig. 10).

Unspecific covering of the transformed cell surface with bovine serum albumin does not lead to the altered growth pattern, and carbohydrates not specific, for the particular agglutinin used do not reverse the change in growth regulation observed.

The specific reversibility, together with the fact that regular growth of transformed cells prior to the confluent density is not inhibited, seems to indicate that the growth inhibition by the trypsin-treated lectin, seen only at high cell densities, is not simply based on a toxic effect.

Both the interpretation of this defect and the mechanism that leads

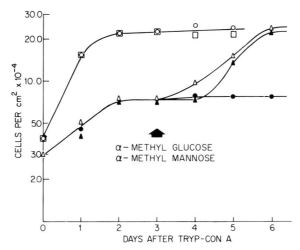

FIG. 10.   Effect of adding monovalent agglutinin and hapten inhibitor to polyoma-transformed 3T3 cells. ○, Py-3T3 cells; ●, Py-3T3 cells + 50 μg/ml monovalent agglutinin; □, Py-3T3 cells + 50 μg/ml monovalent agglutinin + 10^{-2} M α-methyl-glucoside; △, Py-3T3 cells + 50 μg/ml monovalent agglutinin + 10^{-2} M α-methyl glucoside or α-methylmannoside added at day 3; ▲, Py-3T3 cells + 50 μg/ml mono-valent agglutinin + 10^{-3} M α-methylglucoside added at day 3. From Burger and Noonan (1970).

to density-dependent inhibition of growth remain speculative. At the present time, one would like to suggest as a working hypothesis that transformed cells have a series of exposed surface sites, among them ag-glutinin receptor sites, and that this exposure is instrumental in bringing about the loss of growth control while the covered state, regardless of whether it is achieved with extraneous plant proteins or intrinsic mem-brane proteins would make the cells susceptible to normal growth control.

Return to growth control after addition of the trypsin-treated lectin may simply be due to a block in the uptake of small molecular nutrients by physically covering the transport machinery. This is not very likely since at least one small molecule (thymidine) is taken up at the same rate for another 24 hours after addition of the trypsinized agglutinin. Another possibility that comes to mind consists of a block of serum factor uptake since binding of serum factors seems to be necessary for growth (Paul et al., 1971) and can also give rise to escape from density-depen-dent growth control (Holley and Kiernan, 1971).

K. D. Noonan has shown that Py-3T3 cells which had stopped growing at confluency after being treated with trypsinized agglutinin will respond, just as do normal 3T3 cells, with a temporary escape from growth control after addition of high amounts of serum. If the trypsinized agglutinin does not come off from the cell surface, which still remains to be proved, then it is not very likely that the mechanism by which trypsinized agglutinin acts on transformed cell cultures would be through direct competition with overgrowth-stimulating serum factors. K. D. Noonan has recently shown that two other cell lines, besides Py-3T3 cells, respond with lower saturation densities to the presence of trypsinized agglutinin.

Bürk (1968) and also Puck's (Hsie and Puck, 1971) and Pastan's (Johnson et al., 1971) group and Sheppard (1971) have shown that $3':5'$-cyclic AMP (cAMP) and N^6,O^{21}-dibutyryl cyclic AMP (Bu$_2$cAMP) can influence the morphology and growth pattern of transformed cells. Sheppard (1971) and Robert Remo (R. Remo and M. M. Burger, unpublished) have found that in the presence of Bu$_2$cAMP, Py-3T3 cells grow to confluency and stop at that point similarly as do cells treated with trypsinized agglutinin (Burger and Noonan, 1970).

Sheppard (1971) has also shown that Py-3T3 cells treated with Bu$_2$cAMP and theophylline temporarily revert to the nonagglutinable state and regain their former growth and agglutinability patterns after removal of the two agents. This was confirmed for Bu$_2$cAMP alone (Hsie et al., 1971). Higher levels of cAMP (Otten et al., 1971) as well as cAMP synthetase (cyclase) (Makman, 1971) were recently reported for growing, and particularly stationary untransformed cells as compared to transformed cells. Our preliminary results confirm these findings on the whole (B. Breckenridge and M. M. Burger, unpublished observations). If cAMP is really significant for normal growth control, then the protease induced escape from density-dependent inhibition of growth should theoretically be reduced or suppressed by Bu$_2$cAMP. B. Bombik could indeed show that 5×10^{-5} M Bu$_2$cAMP completely quenched both a protease-induced as well as a serum-induced escape from growth control (B. Bombik and M. M. Burger, unpublished). This can not yet be taken as evidence for a role of cAMP in density-dependent inhibition of growth, since this agent may simply prevent, e.g., the metabolic or movement machinery from responding to the surface trigger set by the protease. It suggests to us, however—more than the data on morphological changes after Bu$_2$cAMP addition since we are working with normal contact-inhibited cells—that cAMP may be considered to be a possible mediator of contact inhibition of growth in untransformed cells. For the present time this should not be considered to be more than a tempting working hypothesis.

It may be that trypsinized agglutinin raises the internal cAMP level by activating the membrane-located adenylcyclase or by inhibiting some of the phosphodiesterases or by any other means (Fig. 11) and that cAMP could be the mediator for density-dependent inhibition of growth

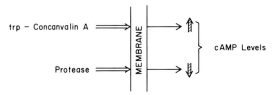

FIG. 11. Hypothetical mechanism of action for the proteases that release untransformed cells from contact inhibition of growth and the action of trypsinized concanavalin A on the return to contact inhibition of growth by transformed cells. From Burger and Noonan (1970).

in untransformed cells. Since cAMP appears to be an almost ubiquitous messenger in an unusually large series of cellular reactions, its specific action in the case of density-dependent growth control will require carefully evaluated evidence to become convincing.

3. The Open Questions. First: Even though the proteolytic treatment necessary for bringing about escape from growth control in untransformed cells is very mild, and just about the same conditions lead to the agglutinable state, the two may be a coincidence, and we are aware that more than only correlative evidence has to be shown before a causal relationship becomes a reasonable assumption. Second: The mechanism through which trypsinized agglutinin can lead phenotypically to normal growth control in transformed cells is entirely open. The possibility that minor amounts of native bivalent agglutinin molecules are still remaining in this trypsinized agglutinin preparation and that they are primarily responsible for the biological effect is unlikely for various reasons but has not yet been rigorously excluded. The monovalency of the trypsinized agglutinin remains to be established by equilibrium dialysis and a quantitative determination of the number of sites per molecule.

III. MITOSIS AND SURFACE CHANGES DURING THE CELL CYCLE OF UNTRANSFORMED CELLS IN CULTURE

A. *Observations and Mitosis*

If the surface membrane could regulate a given part of the whole cell cycle in one way or another, one would expect it to show some cyclic alterations that would be repeated once in every generation. The fact that an active search for such surface changes has only recently been started is largely due to the scarcity of traits or membrane markers that could be examined. Kuhns and Bramson (1968) reported that "H" blood group activity in HeLa cells can be expressed during mitotic waves.

We have found a change in the surface structure during mitosis as detected with a fluorescent agglutinin (Fox *et al.*, 1971). Cikes and Friberg (1971) have shown that H-2 antigen and Moloney leukemia virus determined surface antigens are expressed primarily in the G_1 phase of the cell cycle.

We detected the increased adsorption of fluorescein-labeled WGA to mitotic cells during a study aimed at the development of a fluorescence assay for tumor cells in human specimens. In a first step, a generally increased binding of fluorescent WGA to virally transformed cells in culture was observed. The same observation was independently made by Biddle *et al.* (1970), who found that fluorescent WGA adsorption in untransformed cultures increased with age. The problem we were coping with was the heterogeneity of staining of untransformed cells right after plating. It turned out that, in healthy cultures, almost exclusively only the cells that were in mitosis showed the typical surface fluorescence seen with transformed cells (Fox *et al.*, 1971). In order better to visualize the nuclei in division, we fixed the cells partially, thereby having some labeled material penetrate the cell and sparing the nuclei in division. Without this partial fixation, adsorption of fluorescent WGA could be abolished with the specific carbohydrate haptens, an indication that adsorption took place at the same specific carbohydrate receptor sites as was found for transformed cells. In synchronized cells, the mitotic index and the fluorescent index coincided (Fig. 12).

Increased fluorescence was not simply due to a crinkling of the surface during rounding off in mitosis since interphase cells, rounded off with EDTA, did not increase their surface fluorescence. Furthermore, cells that are fully spread out in early G_1 (still closely together and recognizable as early G_1 cells) were quite strongly fluorescent.

Fluorescent WGA adsorption was earliest observed in late prophase, essentially never at the onset of prophase, and seemed to be shifted by the same delay interval into early G_1.

B. The Open Questions

The most intriguing question here is that of the significance of this observation. Is this surface alteration just a consequence of the various metabolic and structural changes occurring during mitosis, or could it be required for later steps in the cell cycle as a mediator in positive control loops which keep the cell cycle going? The nucleus may, for example, go into the next replication of chromosomal material only when it knows that not only the genetic material has been divided into two parts, but these two sets of chromosomes are located in two new cells, a process that is not terminated until the surface membrane has divided the two cytoplasms with their respective nuclei into two.

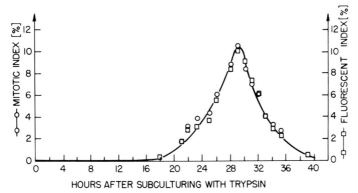

FIG. 12. Mitotic and fluorescence indices of synchronized 3T3 cells. Cells were synchronized by trypsinizing a confluent 3-day-old monolayer of 3T3 cells and replating at lower densities on coverslips. The maximum percentage of synchrony obtained (12–15%) is approximately that expected from other techniques. Coverslips were exposed to fluorescein isothiocyanate-labeled agglutinin, fixed with ethanol, stained with Evans blue and mounted in Elvanol. In control experiments, cells were exposed to fluorescein-labeled agglutinin and counted without fixing and staining, and fluorescence indices were identical with those reported in the figure. Blind counts of several hundred cells were made by two investigators and were in good agreement. From Fox *et al.* (1971).

We are presently investigating whether the surface alteration may therefore not occur just prior to all those situations where the cell goes into its resting states (G_1 or G_0), for example, untransformed cells at confluency.

The observation that mitotic cells display only temporarily the same surface alteration as do transformed cells throughout interphase suggested to us the following working hypothesis, shown in Fig. 13. Infection at any point during the cell cycle may give rise to an abortive mitotic surface change or, more likely, nothing might happen until the next mito-

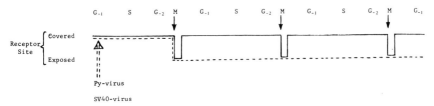

FIG. 13. Hypothetical model for the exposure of the surface configuration typical for transformed cells after infection. In some cases, partial covering may occur after the first mitosis subsequent to infection. In these cases, the transformed surface configuration would be achieved permanently only after the second or third interphase.

sis leads to its usual surface alteration. Infection will prevent regeneration of the usual interphase surface configuration, when the infected cell will return to the G_1 phase, unless the cell was not abortively transformed but permanently transformed, it will remain with its surface in a perpetually "frozen" mitotic surface configuration, a fact which may have important consequences for growth control of all tumor cells.

ACKNOWLEDGMENTS

This work was supported in part by PHS grants CA-10151, 1-K4-CA-16765, SVCP-NCI Contract 71-2372, and temporarily by E. R. Squibb and Sons, Inc. and the Anita G. Mestres Fund.

REFERENCES

AARONSON, S. A., and TODARO, G. (1968). Basis for the acquisition of malignant potential by mouse cells cultured in vitro. Science 162, 1024–1026.

AMBROSE, E. G., DUDGEON, I. A., EASTY, D. M., and EASTY, G. C. (1961). The inhibition of tumor growth by enzymes in tissue culture. Exp. Cell Res. 24, 220–227.

AUB, J. C., TIESLAW, C., and LANKESTER, A. (1963). Reaction of normal and tumor cell surfaces to enzymes. I. Wheat-germ lipase and associated mucopolysaccharides. Proc. Nat. Acad. Sci. U.S. 50, 613–619.

AUB, J. C., SANFORD, B. H., and COTES, M. N. (1965a). Studies on reactivity of tumor and normal cells to a wheat germ agglutinin. Proc. Nat. Acad. Sci. U.S. 54, 396–399.

AUB, J. C., SANFORD, B. H., and WANG, L. H. (1965b). Reactions of normal and leukemia cell surfaces to a wheat germ agglutinin. Proc. Nat. Acad. Sci. U.S. 54, 400–402.

BENJAMIN, T. L. (1970). Host range mutants of polyoma virus. Proc. Nat. Acad. Sci. U.S. 67, 394–399.

BENJAMIN, T. L., and BURGER, M. M. (1970). Absence of a cell membrane alteration function in nontransforming mutants of polyoma virus. Proc. Nat. Acad. Sci. U.S. 67, 929–934.

BIDDLE, F., CRONIN, A. P., and SANDERS, F. K. (1970). Interaction between wheat germ agglutinin and receptors on normal and transformed cells and on erythrocytes. Cytobios 2, 9–17.

BOMBIK, B., and BURGER, M. M. (1973). cAMP and the cell cycle. Inhibition of growth stimulation. Exp. Cell Res. 80, 88–94.

BOREK, C., GROB, M., and BURGER, M. M. (1973). Surface alterations in transformed epithelial and fibroblastic cells in culture: A disturbance of membrane degradation versus biosynthesis. Exp. Cell Res. 77, 207–215.

BURGER, M. M. (1969). A difference in the architecture of the surface membrane of normal and virally transformed cells. Proc. Nat. Acad. Sci. U.S. 62, 994–1001.

BURGER, M. M. (1970a). Changes in the chemical architecture of transformed cell surfaces. In "Permeability and Function of Biological Membranes" (L. Bolis, ed.), pp. 107–119. North-Holland Publ., Amsterdam.

BURGER, M. M. (1970b). Proteolytic enzymes initiating cell division and escape from contact inhibition of growth. Nature (London) 227, 170–171.

BURGER, M. M. (1971a). Forsman antigen exposed on surface membrane after viral transformation. *Nature (London)* **231**, 125–126.

BURGER, M. M. (1971b). The significance of surface structure changes for growth control under crowded conditions. *Growth Contr. Cell Cult., Ciba Found. Symp., 1970* pp. 45–69.

BURGER, M. M., and GOLDBERG, A. R. (1967). Identification of a tumor-specific determinant on neoplasic cell surfaces. *Proc. Nat. Acad. Sci. U.S.* **57**, 359–366.

BURGER, M. M., and JANSONS, V. K. (1972). Complex carbohydrates in transformed cell surfaces as indicators for functional changes. *In* "Glycoproteins of Blood Cells and Plasma" (G. A. Jamieson, ed.), pp. 267–279. Lippincott, Philadelphia, Pennsylvania.

BURGER, M. M., and NOONAN, K. D. (1970). Restoration of normal growth by covering of agglutinin sites on tumor cell surface. *Nature (London)* **228**, 512–515.

BÜRK, R. R. (1968). Reduced adenyl cyclase activity in a polyoma virus transformed cell line. *Nature (London)* **219**, 1272–1275.

CIKES, M., and FRIBERG, S. (1971). Expression of H-2 and Maloney leukemia virus-determined cell surface antigens in synchronized cultures of a mouse cell line. *Proc. Nat. Acad. Sci. U.S.* **68**, 566–569.

ECKHART, W., DULBECCO, R., and BURGER, M. M. (1971). Temperature-dependent surface changes in cells infected or transformed by a thermosensitive mutant of polyoma virus. *Proc. Nat. Acad. Sci. U.S.* **68**, 283–286.

FOX, T., SHEPPARD, J. R., and BURGER, M. M. (1971). Cyclic membrane changes in animal cells: Transformed cells permanently display a surface architecture detected in normal cells only during mitosis. *Proc. Nat. Acad. Sci. U.S.* **68**, 244–247.

GEPNER, I., and STEINBERG, M. (1973). Are conconavalin A receptor sites mediators of cell-cell adhesion? *Nature (London), New Biol.* **241**, 249–251.

HAKOMORI, S-I., KOSCIELAK, J., BLOCH, K. J., and JEANLOZ, R. W. (1967). Immunologic relationship between blood group substances and a fucose-containing glucolipid of human adenocarcinoma. *J. Immunol.* **98**, 31–38.

HÄYRY, P., and DEFENDI, V. (1970). Surface antigen(s) of SV40-transformed tumor cells. *Virology* **41**, 22–29.

HOLLEY, R. W., and KIERNAN, J. A. (1971). Studies of serum factors required by 3T3 and SV3T3 cells. *Growth Contr. Cell Cult., Ciba Found. Symp., 1970* pp. 3–15.

HSIE, A. W., and PUCK, T. T. (1971). Morphological transformation of Chinese hamster cells by dibutyryl adenosine cyclic 3′:5′-monophosphate and testosterone. *Proc. Nat. Acad. Sci. U.S.* **68**, 358–361.

HSIE, A. W., JONES, C., and PUCK, T. T. (1971). Further changes in the differentiation state accompanying the conversion of Chinese hamster cells to fibroblastic form by dibutyryl adenosine cyclic 3′:5′-monophosphate and hormones. *Proc. Nat. Acad. Sci. U.S.* **68**, 1648–1652.

INBAR, M., and SACHS, L. (1969a). Interaction of the carbohydrate-binding protein conconavalin A with normal and transformed cells. *Proc. Nat. Acad. Sci. U.S.* **63**, 1418–1425.

INBAR, M., and SACHS, L. (1969b). Structural differences in sites on the surface membrane of normal and transformed cells. *Nature (London)* **223**, 710.

INBAR, M., RABINOWITZ, Z., and SACHS, L. (1969). The formation of variants with a reversion of properties of transformed cells. III. Reversion of the structure of the cell surface membrane. *Int. J. Cancer* **4**, 690–696.

Jacob, F., Brenner, S., and Cusin, F. (1963). On the regulation of DNA replication in bacteria. *Cold Spring Harbor Symp. Quant. Biol.* **27**, 329–348.

Johnson, G. S., Friedman, R. M., and Pastan, I. (1971). Restoration of several morphological characteristics of normal fibroblasts in sarcoma cells treated with adenosine-3′:5′-cyclic monophosphate and its derivatives. *Proc. Nat. Acad. Sci. U.S.* **68**, 425–429.

Kuhns, W. J., and Bramson, S. (1968). Variable behavior of blood group H on HeLa cell populations synchronized with thymidine. *Nature (London)* **219**, 938–939.

Makman, M. H. (1971). Conditions leading to enhanced response to glucagon, epinephrine, or prostaglandins by adenylate cyclase of normal and malignant cultured cells. *Proc. Nat. Acad. Sci. U.S.* **68**, 2127–2130.

Nägeli, C., and Cramer, C. (1855). "Pfanzenphysiologische Untersuchungen," No. 1. Schulthess, Zürich.

Onodera, K., and Sheinin, R. (1970). Macromolecular glucosamine-containing component of the surface of cultured mouse cells. *J. Cell Sci.* **7**, 337–355.

Otten, J., Johnson, G. S., and Pastan, I. (1971). Cyclic AMP levels in fibroblasts: Relationship to growth rate and contact inhibition of growth. *Biochem. Biophys. Res. Commun.* **44**, 1192–1198.

Ozanne, B., and Sambrook, J. (1971). Isolation of lines of cells resistant to agglutination by conconavalin A from 3T3 cells transformed by SV40. *In* "The Biology of Oncogenic Viruses" (G. Silvestri, ed.), pp. 248–257. North-Holland Publ., Amsterdam.

Paul, D., Lipton, A., and Klinger, I., (1971). Serum factor requirements of normal and simian virus 40 transformed 3T3 mouse fibroblasts. *Proc. Nat. Acad. Sci. U.S.* **68**, 645–648.

Pollack, R. E., and Burger, M. M. (1969). Surface specific characteristics of a contact-inhibited cell line containing the SV40 viral genome. *Proc. Nat. Acad. Sci. U.S.* **62**, 1074–1076.

Pollack, R. E., Green, H., and Todaro, G. (1968). Growth control in cultured cells: Selection of sublines with increased sensitivity to contact inhibition and decreased tumor-producing ability. *Proc. Nat. Acad. Sci. U.S.* **60**, 126–133.

Sanford, K. K. Barker, B. E., Woods, M. W., Parshad, R., and Law, L. W. (1967). Search for "indications" of neoplastic conversion *in vitro*. *J. Nat. Cancer Inst.* **39**, 705–733.

Sefton, B. M., and Rubin, H. (1970). Rtlease from density dependent growth inhibition by proteolytic enzymes. *Nature (London)* **227**, 843–845.

Sela, B. A., Lis, H., Sharon, N., and Sachs, L. (1970). Different locations of carbohydrate-containing sites in the surface membrane of normal and transformed mammalian cells. *J. Membrane Biol.* **3**, 267–279.

Sheinin, R. (1972). Studies on the surface moieties of normal and virus-transformed mouse fibroblasts. *In* "Cell Differentiation" (R. Harris, P. Allin, and D. Viza, eds.), pp. 186–190. Munksgaard, Copenhagen.

Sheppard, J. R. (1971). Restoration of contact inhibited growth to transformed cells by dibutyryl adenosine 3′:5′-cyclic monophosphate. *Proc. Nat. Acad. Sci. U.S.* **68**, 1316–1320.

Shier, W. I. (1971). Preparation of a "chemical vaccine" against tumor progression. *Proc. Nat. Acad. Sci. U.S.* **68**, 2078–2082.

Vasiliev, J. M., Gelfand, I. M., Guelstein, V. I., and Fetisova, E. K. (1970). Stimulation of DNA synthesis in cultures of mouse embryo fibroblast-like cells. *J. Cell. Physiol.* **75**, 305–313.

Epidermal Growth Factor:
Chemical and Biological Characterization[1]

STANLEY COHEN AND JOHN M. TAYLOR

Department of Biochemistry, Vanderbilt University, Nashville, Tennessee

I. INTRODUCTION

During the course of studies on the nerve growth-promoting protein of the submaxillary gland of the male mouse (Cohen, 1960), it was observed that partially purified extracts of these glands caused gross anatomical changes when injected daily into newborn mice. Among the specific tissue changes observed were the precocious opening of the eyelids (6–7 days instead of the normal 12–14 days), and the precocious eruption of the incisors (6–7 days instead of the normal 9–10 days). By means of a biological assay based on the precocious opening of the eyelids of the newborn mouse, a polypeptide responsible for this effect was isolated (Cohen, 1962). The biological activity was subsequently found to be due to a direct stimulation of the proliferation and keratinization of epidermal tissue; therefore, the polypeptide has been termed epidermal growth factor (EGF). The chemical and physical properties of EGF will be described, followed by a discussion of its biological activity.

[1] Much of this work was supported by USPHS Grants HD 00700 and FR 06067.

II. CHEMISTRY OF THE EPIDERMAL GROWTH FACTOR

A. Early Characterization

The isolation procedure for EGF involved standard methods of protein purification. In a typical preparation, the submaxillary glands of 150 adult male mice, having a wet weight of approximately 22–25 gm, yielded 5–7 mg of pure EGF, with a recovery of approximately 20% of the original activity. EGF was estimated to be approximately 0.5% of the dry weight of the submaxillary gland protein.

The isolated material appeared to be relatively pure and homogeneous by several different criteria. When it was examined in the analytical ultracentrifuge, EGF showed only a single component having a sedimentation coefficient $(s_{20,w})$ of 1.25 S. Paper electrophoresis in buffers of different pH, and paper chromatography in a variety of solvents, showed only one spot after staining with protein dyes. The material was antigenic; and cellulose acetate immunoelectrophoresis of the factor against its antiserum revealed only one precipitin band.

The ultraviolet absorption spectrum of the purified factor was observed to be typical for proteins, with a 280:260 optical density ratio of 1.6. EGF was found to be heat stable and nondialyzable. Its biological activity was stable to boiling in water but was destroyed by heating in dilute alkali or dilute acid. Biological activity was also destroyed by incubation with chymotrypsin or a bacterial protease, but was only partially lost after incubation with trypsin. These data strongly suggested that the factor was polypeptide-like in nature (Cohen, 1962).

B. Chemical and Physical Properties

A more detailed study of the chemical and physical properties of EGF has been undertaken (Taylor et al., 1970; J. M. Taylor, S. Cohen, and W. M. Mitchell, unpublished observations). Studies of the amino acid composition, summarized in Table 1, indicate the absence of three specific amino acid residues: lysine, alanine, and phenylalanine. The calculated number of residues was based on one histidine residue per mole of polypeptide. The presence of six half-cystine residues has been confirmed by amino acid analysis of the S-aminoethylated derivative of EGF. No free sulfhydryl groups were detected, an indication of the probable existence of three disulfide bridges within the molecule. In addition, no detectable hexosamines have been found.

TABLE 1

AMINO ACID COMPOSITION OF LOW MOLECULAR WEIGHT (MW) EGF[a]

Amino acid	Residues per mole	Amino acid	Residues per mole
Lysine	0	Alanine	0
Histidine	1	Half-cystine	6
Arginine	4	Valine	2
Aspartic acid	8–9	Methionine	1
Threonine	2	Isoleucine	2
Serine	6–7	Luecine	4
Glutamic acid	3–4	Tyrosine	5
Proline	2	Phenylalanine	0
Glycine	6–7	Tryptophan	2
		Total residues	54–58
		Minimum MW	6166–6554

[a] Samples of 100 μg were hydrolyzed in 6 N HCl under vacuum for 24 hours. The average value from two hydrolyses is presented. The calculated number of residues was based on one histidine and four arginine residues per mole of polypeptide. The results were uncorrected for hydrolytic losses. The recovered amino acids amounted to 85% by weight of the starting material. Tryptophan was measured spectrophotometrically. From Taylor et al. (1970).

A minimum molecular weight of 6166–6554 was calculated from the amino acid composition. This molecular size is supported by low speed sedimentation equilibrium studies, which indicate a molecular weight of 6400, and by gel filtration studies on columns of BioGel P-10, which suggest an approximate molecular weight of 7000.

The values for the molecular weight of EGF reported here differ from the original estimate of 15,000 (Cohen, 1962). The earlier estimate, however, was calculated from the amino acid composition on the basis of one alanine and ten leucine residues per mole. The recent amino acid analyses of EGF, prepared under improved conditions which reduce contamination, do not reveal the presence of a significant amount of alanine.

Studies with chemically modified derivatives of EGF indicate that it is a single-chain polypeptide. This conclusion is supported by the finding of only one amino-terminal residue (asparagine) and only one carboxyl-terminal residue (arginine).

The results of these and other investigations on the chemical and physical properties of EGF are summarized in Table 2.

C. High Molecular Weight Form of Epidermal Growth Factor

In crude homogenates of the submaxillary glands of adult male mice,

TABLE 2
PHYSICAL AND CHEMICAL PROPERTIES
OF EPIDERMAL GROWTH FACTOR

Property	Value
Molecular weight in daltons	
Sedimentation equilibrium	6400
Gel filtration (BioGel P-10)	7000
Amino acid composition	6166–6554
Sedimentation coefficient ($s_{20,w}$)	1.25 S
Partial specific volume ($\bar{v}$) in cm^3/gm	0.69
Extinction coefficient ($E_{1\ cm,\ 280nm}^{1\ \%}$)	30.9
Isoelectric point	pH 4.60
Conformation	Nonhelical
Number of polypeptide chains	One
Amino terminus	Asparagine
Carboxyl terminus	Arginine
Disulfide bonds	Three
Missing amino acids	Lys, Ala, Phe
Hexosamine content	None detected

EGF was found to be a component of a high molecular weight complex (Taylor *et al.*, 1970). The isolation and characterization of this biologically active complex, termed high molecular weight epidermal growth factor (HMW-EGF), will be described below.

The existence of HMW-EGF was first detected by the gel filtration of aqueous extracts of the submaxillary glands on calibrated columns of Sephadex G-75. The result of a gel filtration experiment is shown in Fig. 1. The eluted fractions were examined by an immunoprecipitation reaction with an antibody to the low molecular weight EGF. The major EGF-antibody precipitating material was observed with the pigmented hemoglobin peak in the 50,000 to 70,000 molecular weight fraction of the eluate. This fraction was also found to be biologically active by means of the eyelid-opening assay in the newborn mouse.

The HMW-EGF was isolated by means of standard methods of protein purification. On the basis of purification yields, HMW-EGF was estimated to be approximately 2–3% of the dry weight of the submaxillary gland protein, with a recovery of approximately two-thirds of the original material.

High speed sedimentation equilibrium studies indicate that HMW-EGF has a molecular weight of approximately 74,000. It can be reversibly dissociated, under a variety of conditions, into two molecules of the low molecular weight EGF (6400), and two molecules of an EGF-binding protein (30,000). Conditions for dissociation include adsorption

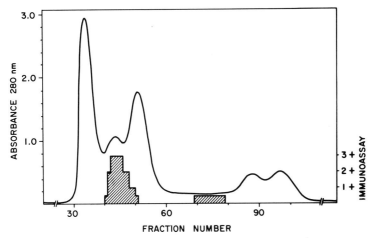

FIG. 1. Gel filtration of a crude extract of male mouse submaxillary glands on Sephadex G-75. A 4.0-ml extract containing 0.3 gm of protein was applied to a Sephadex G-75 column (2.5 × 90 cm) equilibrated with the elution buffer (0.01 M sodium acetate, pH 5.9, and 0.1 M sodium chloride). The flow rate was 3.8 ml/cm² per hour, and 5.0-ml fractions were collected. The hatched areas indicate the immunoprecipitation reaction of the eluate fractions with the antibody to low molecular weight epidermal growth factor. The immunoassay was carried out by layering 25 μl from each eluate fraction over 25 μl of antiserum and qualitatively observing the precipitate that formed at the interface after 30 minutes. From Taylor (1970).

to ion-exchange columns, gel filtration in buffers below pH 5.0 and above pH 8.0, and by isoelectric focusing with low pH range ampholyte solutions. The result of an isoelectric focusing experiment is shown in Fig. 2. Only two major protein peaks are observed above the background absorbance. The peak at pH 4.60 corresponds to low molecular weight EGF, and the peak at pH 5.60 corresponds to the EGF-binding protein. The low molecular weight EGF obtained from the dissociation of HMW-EGF is identical to the EGF as originally isolated by Cohen (1962). Criteria of identity include isoelectric point determinations, amino acid composition, carboxyl-terminal residue analysis, gel filtration studies, and antigenic identity with an antibody prepared against EGF as originally isolated.

The low molecular weight EGF and the EGF-binding protein can be recombined to form a high molecular weight complex, having approximately the same molecular weight as the native HMW-EGF when examined by sedimentation equilibrium and gel filtration studies. In a typical experiment, equal weights of EGF and the EGF-binding protein are mixed together in a neutral buffer, allowed to stand for a few hours, then

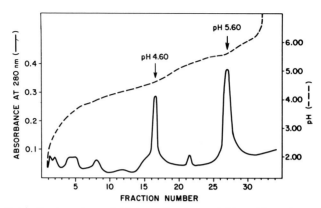

FIG. 2. Subunit character of high molecular weight epidermal growth factor (HMW-EGF) by isoelectric focusing under dissociation conditions. The pH range of the ampholyte solution was from pH 4 to 6. The solid line shows the absorbance at 280 nm of the fractions (3 ml) obtained from the isoelectric focusing column. The dashed line represents the pH gradient developed during the experiment. The experiment was performed with 8 mg of HMW-EGF. From Taylor *et al.* (1970).

applied to a calibrated column of Sephadex G-100. The results of this experiment are shown in Fig. 3. The EGF-binding protein alone did not form a precipitate with the EGF antibody, whereas the recombined HMW-EGF did form a precipitate. In addition, the elution volume of the recombined HMW-EGF was similar to that of the native HMW-EGF, and distinctly different from the elution volumes of its two components.

After the HMW-EGF had been isolated, the observation was made that the EGF-binding protein was an arginine esterase, and various enzymatic parameters were measured. The esterase showed a hydrolytic specificity for arginine esters, and for lysine esters to a lesser degree. Significant catalytic hydrolysis was not observed for a variety of other amino acid esters, with no detectable hydrolysis of arginine-amide substrates. The rate of hydrolysis of benzoyl-arginine ethyl ester at 25°C was approximately 390 μmoles per minute per mg of enzyme at pH 8.0. The additional observation that EGF possesses a carboxyl-terminal argine residue suggests that EGF may be generated from a precursor protein by the possible proteolytic action of the EGF-binding esterase.

In a broad context, the observations, that certain other polypeptide hormones, such as bradykinin (Schachter, 1969) and insulin (Chance *et al.*, 1968), arise from precursors by the proteolytic action of arginine esterases, suggest that the formation of active polypeptide hormones from

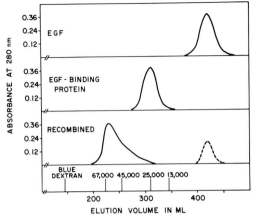

Fig. 3. Recombination of epidermal growth factor (EGF) binding protein and low molecular weight epidermal growth factor. Samples containing 8–16 mg of protein in 4 ml of elution buffer (0.01 M sodium acetate, pH 5.9, and 0.1 M sodium chloride) were applied to a Sephadex G-100 column (2.5 × 90 cm) equilibrated with the same buffer. The flow rate was 4.3 ml/cm² per hour, and 5-ml fractions were collected. The top curve shows the elution profile of pure low molecular weight EGF. The middle curve shows the elution profile of pure EGF-binding protein. The bottom curve shows the elution profile of the recombined high molecular weight EGF (the dotted line indicates excess unreacted low molecular weight EGF). At the bottom of the figure is shown the elution volumes and molecular weights of various protein standards: bovine serum albumin (67,000), ovalbumin (45,000), chymotrypsinogen A (25,000), and horse heart cytochrome c (13,000). From Taylor et al. (1970).

inactive precursors, by the proteolytic action of a family of arginine esterases, may be a general phenomenon.

It may be noted that the nerve growth factor of the male mouse submaxillary gland, originally isolated by Cohen (1960), has been found to be a component of a high molecular weight complex, with an approximate molecular weight of 140,000 (Varon et al., 1967). One of the subunits of this complex has been found to be an arginine esterase (Greene et al., 1969). The physical and enzymatic properties of this enzyme bear a close resemblance to the EGF-binding esterase. Preliminary immunological evidence indicates that these two enzymes may be nearly identical, raising the possibility that the biosynthesis and activation of the epidermal growth factor and the nerve growth factor may be quite similar, perhaps under the control of the same genetic locus. This possibility is also suggested by their similar sexually dimorphic character and sensitivity to hormonal induction by testosterone.

III. BIOLOGY OF THE EPIDERMAL GROWTH FACTOR

A. Effects of EGF in Vivo

The precocious opening of the eyelids and incisor eruption following the daily subcutaneous injection of microgram quantities of EGF is ascribed mainly to an enhancement of epidermal growth and keratinization. A histological examination of these animals revealed enhanced keratinization and an increase in the thickness of the epidermis not only in the eyelid area (Fig. 4), but also in the back skin and epithelium lining the mouth (Cohen and Elliott, 1963). This histological picture has been confirmed by making a number of chemical measurements (protein, RNA, DNA) of pure epidermis obtained by trypsinization of standard areas of skin from 5-day-old control and EGF-treated rats (Angeletti *et al.*, 1964; Mann and Fenton, 1970). Farebrother and Mann (1970) have extended these histological observations. The most noticeable effects again were seen in the skin. Changes were also noted in the pericardium, kidney capsule, and bile duct. A diminution of the thickness and fat content of the dermis was reported. An enhancement of the carcinogenicity of topically applied 3-methylcholanthrene on mouse skin has also been noted (Reynolds *et al.*, 1965) in EGF-treated animals.

The eyelid opening effect in newborn rats is demonstrable at a dosage level of 0.1 μg/gm per day. At higher dosage levels, 1–2 μg/gm per day,

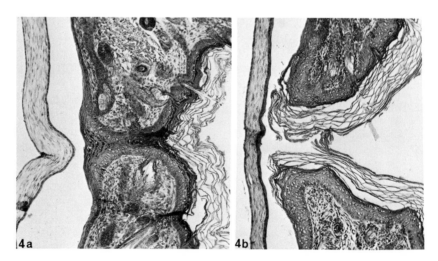

Fig. 4. Cross sections of the eyelid area from (a) control and (b) experimental 8-day-old rats. The experimental animal had received daily injections (1 μg per gram body weight) of EGF. From Cohen and Elliott (1963).

there is a distinct growth inhibition of the animals, and after 10 days of treatment, clear morphological changes are visible in the liver, where large accumulations of fat are present. This fatty liver appears to be, almost exclusively, the result of the net accumulation of triglycerides (Heimberg *et al.*, 1965).

B. Effects of EGF on Cells in Culture

EGF has been shown to stimulate epithelial cell proliferation in a number of organ culture systems. The initial observations were made with fragments of skin from the trunk of 7-day chick embryos, cultured in synthetic medium. The number of layers of epidermal cells of control cultures remained almost unchanged during a 48-hour incubation period whereas a marked increase in the number of layers of epidermal cells occurred in the experimental cultures (Fig. 5). In the constant presence

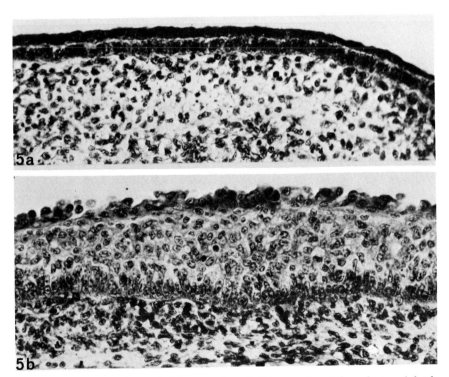

FIG. 5. Cross sections of (a) control and (b) experimental explants of back skin from 7-day chick embryos after 3 days of incubation. The experimental cultures contained 5 μg of EGF per milliliter. From Cohen (1965a).

of [³H]thymidine, almost every basal cell of the EGF-treated cultures became labeled, while in control cultures only very few basal cells contained radioactivity (Cohen, 1965a).

This proliferative effect of EGF has also been reported for epithelial cells of mouse mammary glands and mammary carcinomas in organ culture (Turkington, 1969a,b).

One of the initial morphological effects of the factor is a stimulation of cell migration, demonstrated by experiments in which aggregates of chick embryo epidermal cells were incubated in collagen-coated culture dishes for 18 hours. Control cells consistently formed compact flattened colonylike areas while, in contrast, in the presence as little as 0.02 μg per milliliter of the factor, the cells spread into a fibroblastlike network (Cohen, 1965b). Under these conditions EGF did not enhance thymidine incorporation, indicating that DNA synthesis had not been stimulated.

C. Biochemical Studies on the Mechanism of Action of EGF

The stimulation of epidermal proliferation in culture is dependent upon a number of conditions, among which are the age of the embryo from which the skin is explanted, and whether dermal cells are present.

In the following experiments, sheets of pure epidermis derived from the back skin of 9-day embryos were cultured on Millipore filters in a variety of media with or without microgram quantities of EGF. This system was chosen, rather than epidermis or whole skin from younger embryos, to obtain sufficient amounts of tissue for biochemical work and permit studies on the "early" effects of EGF on a single cell type. The observations and conclusions from a number of studies (Hoober and Cohen, 1967a,b; Cohen and Stastny, 1968) are summarized below.

1. In the control tissues no significant changes occurred in the total amounts of protein, RNA, or DNA during the 3-day period of incubation. However, total net protein and RNA in the experimental cultures increased 2-fold over their initial values during the first 48 hours of incubation. No significant increase in the content of DNA in epidermis cultured in the medium containing EGF was observed (Fig. 6). The failure of EGF to stimulate DNA synthesis under these conditions, as well as the clear stimulation of protein and RNA synthesis, has been confirmed in experiments in which the appropriate radioactive precursors (thymidine, orotic acid, uridine, and lysine) were added to the medium in the presence and in the absence of EGF, and their incorporation into the tissues was studied.

2. EGF appears to rapidly stimulate the transport of certain metabolites. Within 15 minutes after the addition of EGF there is an approximately 2-fold stimulation of the uptake of radioactive aminoisobutyric

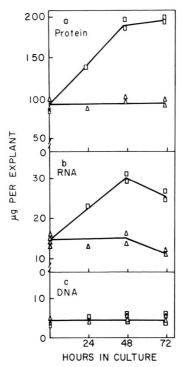

Fig. 6. Changes in (a) total protein, (b) RNA, and (c) DNA of 8.5-day chick embryo epidermis during culture with 5 µg per milliliter of EGF (□) and without EGF (△). From Hoober and Cohen (1967a).

acid and uridine into the trichloroacetic acid soluble fraction of the cells. This uptake is not prevented by inhibitors of protein synthesis such as cycloheximide, indicating that the synthesis of new proteins was not required for these permeability changes.

3. During the first 90 minutes following the addition of EGF there is an approximately 3-fold stimulation of the incorporation of labeled uridine into total RNA. This stimulation is not inhibited by puromycin or cycloheximide.

4. A number of experiments have been performed to partially characterize the nature of the newly formed RNA which appears in the cytoplasm after EGF addition. After a 90-minute incubation period with labeled uridine in the presence and absence of EGF, sucrose sedimentation analysis of the mitochondrial supernatants revealed that the synthesis of all types of RNA discernible on the gradient (4 S, 16 S, 28 S and heterogeneous RNA) were stimulated (approximately 4- to 8-fold).

5.   In the presence of EGF there is a conversion of preexisting ribosomal monomers into functional polysomal structures. (Controls contained approximately 30% of the total ribosomes as polysomes whereas EGF-treated cells contain about 60% as polysomes.) This ribosomal monomer–polysome conversion is detectable on sucrose gradients 30 minutes after the addition of EGF to the culture medium, and is observable even in the presence of cycloheximide or in a simple salt medium. It appears, therefore, that the initial transfer of monosomes to polysomes does not *require* the synthesis of new protein or an increased transport of amino acids and glucose. EGF also stimulates the synthesis of new ribosomal subunits.

6.   The increase in the ratio of polysomes to monosomes in the epidermal cells under the influence of EGF would suggest that the isolated total ribosomal population of the EGF-treated cells should be more active in cell-free protein synthesis than that of control cells. To examine this question, a cell-free protein-synthesizing system from cultured chick embryo epidermis was devised. Extracts prepared from epidermis cultured with EGF were over 2 times as active as extracts of tissue cultured without EGF in incorporating labeled amino acids (using endogenous messenger RNA). This difference was accounted for in the ribosomal fraction of the cells, with no differences detected in the soluble fraction. In addition, the total ribosomal population from treated cells was more active in protein synthesis even in the presence of polyuridylic acid than were control ribosomes. This suggested that ribosomes from treated cells had a greater ability to bind messenger RNA in a functional manner than those of control cells. These differences were detectable 30 minutes after the addition of EGF.

We were led to the next series of experiments (Stastny and Cohen, 1970) by the similarities between the response of epidermal cells to EGF and the response of many other cell types to hormonal growth stimulation, and to the results of several groups on the induction of ornithine decarboxylase in liver by partial hepatectomy or growth hormone administration (Jänne *et al.*, 1968; Russell and Snyder, 1968; Jänne and Raina, 1969).

Briefly, EGF induces a marked (40-fold), but transient, increase of ornithine decarboxylase activity in cultures of chick embryo epidermis (Fig. 7). The induction of the enzyme is prevented by inhibitors of protein synthesis (cycloheximide and puromycin and fluorophenylalanine), suggesting that *de novo* synthesis of the enzyme had occurred. However, until the enzyme is isolated or measured in some absolute manner, this conclusion can only be a tentative one.

The induced ornithine decarboxylase activity was reflected in the intra-

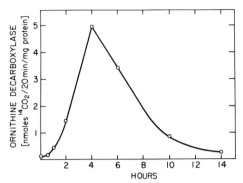

Fig. 7. Time course of the induction of ornithine decarboxylase activity in chick embryo epidermis by EGF. The reaction mixture contained 3.0 μmoles of sodium-potassium phosphate buffer (pH 6.6), 0.1 μmole of EDTA, 12 nmoles of pyridoxal phosphate, 0.12 μmole of DL-[1-^{14}C]ornithine, and approximately 0.3 mg of tissue protein, in a total final volume of 60 μl. From Stastny and Cohen (1970).

cellular accumulation of labeled putrescine in the presence of EGF when labeled ornithine (Fig. 8) or arginine was present in the medium. EGF also induces lysine decarboxylase activity with a concomitant increase in intracellular cadaverine in the presence of labeled lysine.

The possible biological significance of the transient induction of ornithine decarboxylase activity for growth regulation in cultures of chick epidermis is supported by the finding that EGF induces this enzyme in skin when injected into neonatal mice. The additional observation that the enzyme is induced in the testes of these animals suggests that this polypeptide may also affect epithelia in other organs.

Whereas the induction of the enzyme is prevented by a number of inhibitors of protein synthesis, several other metabolic events (discussed previously), which are stimulated by EGF, are not prevented by inhibition of protein synthesis. These results suggest that the induction of ornithine decarboxylase is an early but secondary event in the action of EGF.

Polyamines have been implicated in a large and diverse number of biological processes (Herbst and Bachrach, 1970). For example, they have been shown to stabilize ribosomes and membranes and to stimulate the synthesis of protein, DNA, RNA, and aminoacyl-transfer RNA. In many of these reactions the polyamines may replace, at least partially, a Mg^{2+} requirement, and it has been suggested that polyamines may be more important for protein synthesis *in vivo* than is Mg^{2+} (Hurwitz and Rosano, 1967).

The fact that both steroid (Cohen *et al.*, 1970; Pegg *et al.*, 1970) and

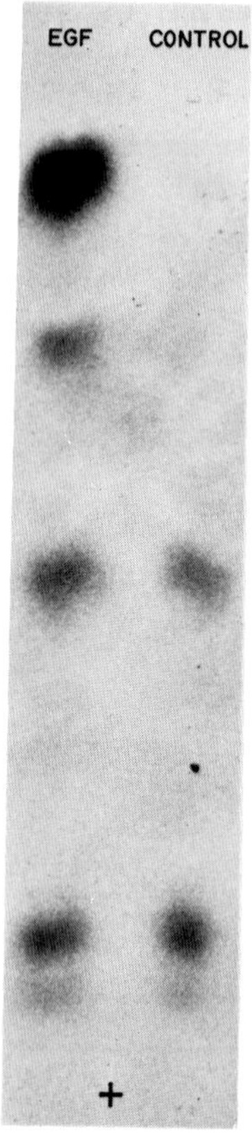

	EGF	CONTROL
		counts / min
PUTRESCINE	573	24
SPERMIDINE	69	<5
SPERMINE	<5	<5
ORNITHINE	114	110

Fig. 8. Intracellular accumulation of polyamines in the presence and in the absence of EGF. Epidermis was cultured in medium containing DL-[2-¹⁴C]ornithine for a 4-hour period. The polyamines were extracted from the tissue and separated by paper electrophoresis, and the labeled components were detected by autoradiographic procedures. From Stastny and Cohen (1970).

polypeptide hormones may act as inducers of ornithine decarboxylase again suggests that the induction of the decarboxylase is an early, but not necessarily primary, event in the action of these hormones. It is conceivable that the rapid induction of ornithine decarboxylase and the subsequent accumulation of putrescine or polyamines provide a mechanism by which the cell may rapidly alter its internal environment to optimize conditions for a variety of biosynthetic reactions.

Thus we have described a series of metabolic alterations which accompany the growth-stimulating effects of EGF on epidermal cells. Many of these changes appear to take place in a variety of cells when a growth stimulus is applied. Neither the initial binding site nor the "primary" metabolic effect of EGF has been clearly identified.

D. Synthesis and Storage of EGF

The tubular cells of the submaxillary gland of rodents exhibit sexual dimorphism. The morphology and granule content of these cells are dependent upon the hormonal status of the animal: the cells are developed fully in the male only after puberty; castration results in the atrophy of the tubular portion of the gland; the injection of testosterone into female mice results in a hypertrophy and hyperplasia of these cells (Sreebny and Meyer, 1964).

The quantity of EGF present in the submaxillary gland closely parallels this development of the tubular system (Cohen, 1965b). A number of other proteins of the submaxillary gland also show this sexually dimorphic character, including nerve growth factor (Cohen, 1960; Levi-Montalcini and Angeletti, 1964), renin (Oliver and Gross, 1967), and certain arginine esterases (Angeletti et al., 1967; Calissano and Angeletti, 1968).

Turkington et al. (1971) have demonstrated by immunofluorescent staining that EGF is present in specific tubular cells of the submaxillary gland, but was not detected in any other mouse tissue examined. Moreover, organ cultures of mouse submaxillary gland incorporated labeled amino acids into a protein that was shown to be identical to authentic EGF by polyacrylamide gel electrophoresis after purification by specific immunoabsorption. These results indicate that EGF is synthesized in the mouse submaxillary gland and suggest that the elaboration of EGF may be a specific function of tubular cells.

E. Radioimmunoassay of EGF

The development of a very sensitive radioimmunoassay for EGF using a rabbit antiserum and iodinated EGF (Byyny et al., 1971a,b) permits further studies on the physiology of this growth-promoting polypeptide. The assay is sensitive to as little as 30 pg of EGF and is not affected by cross-reactivity to a variety of polypeptide hormones (ACTH, growth hormone, FSH, LH, TSH, or insulin). Using this assay it was possible to monitor quantitatively the accumulation of EGF in the submaxillary gland in male mice of varying age (Fig. 9). The EGF content of the gland was very low in 15-day-old mice, about 0.016 ng/mg wet tissue,

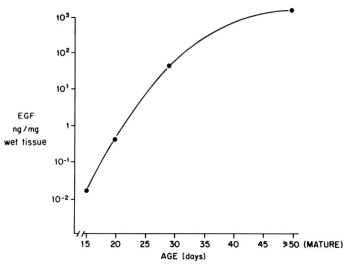

Fig. 9. The content of epidermal growth factor (EGF) in the submaxillary glands of male mice during maturation.

and increased to levels of about 1000 ng/mg in glands of mature male mice.

EGF has been found to be present in mouse serum at a level of approximately 1 ng/ml. The stimulation of α-adrenergic receptors (intravenous injection of phenylephrine at 1.5 μg/gm) in normal adult male mice leads to a marked increase in the serum levels of EGF, reaching 150 ng/ml in 60 minutes. These findings support the view that EGF serves a continuing physiological function in the animal, and that its secretion may be regulated by feedback mechanisms.

Considerable amounts of EGF have been detected in normal mouse milk (approximately 200 ng/ml). In preliminary experiments it has been possible to elicit precocious eyelid opening and incisor eruption by oral administration of EGF to newborn mice. This suggests the possibility that an EGF-like molecule may be present in milk and exert its effects on the neonatal animal.

Although there is at present no direct evidence for the role of the growth factor in normal development and cell control, the following indirect evidence indicates that an important function does exist: EGF was isolated from a mammalian organ; the concentration of the factor in the submaxillary gland is dependent on the hormonal status of the animal; the secretion of EGF into the serum is stimulated by α-adrenergic compounds; and finally, EGF has specific morphological and biochemical effects on epidermis and certain epithelial tissues.

REFERENCES

ANGELETTI, P. U., SALVI, M. L., CHESANOW, R. L., and COHEN, S. (1964). Azione dell' "epidermal growth factor" sulla sintesi di acici nucleici e protine dell'epitelio cutaneo. *Experientia* 20, 146–148.

ANGELETTI, R. A., ANGELETTI, P. U., and CALISSANO, P. (1967). Testosterone induction of estero-proteolytic activity in the mouse submaxillary gland. *Biochim. Biophys. Acta* 139, 372–381.

BYYNY, R. L., ORTH, D. N., COHEN, S., DOYNE, E. S., and ISLAND, D. P. (1971a). Epidermal growth factor radioimmunoassay: Effects of age, androgen, and adrenergic agents on EGF storage and release. *J. Clin. Endocrinol. Metab.* 32, A45.

BYYNY, R. L., ORTH, D. N., COHEN, S., and ISLAND, D. P. (1971b). Solid phase radioimmunoassay (RIA) for epidermal growth factor (EGF). *Clin. Res.* 19, 29.

CALISSANO, P., and ANGELETTI, P. U. (1968). Testosterone effect on the synthetic rate of two esteropeptidases in the mouse submaxillary gland. *Biochim. Biophys. Acta* 156, 51–58.

CHANCE, R. E., ELLIS, R. M., and BROMER, W. W. (1968). Porcine proinsulin: Characterization and amino acid sequence. *Science* 161, 165–167.

COHEN, S. (1960). Purification of a nerve growth promoting protein from the mouse salivary gland and its neuro-cytotoxic antiserum. *Proc. Nat. Acad. Sci. U.S.* 46, 302–311.

COHEN, S. (1962). Isolation of a mouse submaxillary gland protein accelerating incisor eruption and eyelid opening in the newborn animal. *J. Biol. Chem.* 237, 1555–1562.

COHEN, S. (1965a). The stimulation of epidermal proliferation by a specific protein (EGF). *Develop. Biol.* 12, 394–407.

COHEN, S. (1965b). Growth factors and morphogenic induction. *In* "Developmental and Metabolic Control Mechanisms and Neoplasia," pp. 251–272. Williams & Wilkins, Baltimore, Maryland.

COHEN, S., and ELLIOTT, G. A. (1963). The stimulation of epidermal keratinization by a protein isolated from the submaxillary gland of the mouse. *J. Invest. Dermatol.* 40, 1–5.

COHEN, S., and STASTNY, M. (1968). Epidermal growth factor. III. The stimulation of polysome formation in chick embryo epidermis. *Biochim. Biophys. Acta* 166, 427–437.

COHEN, S., O'MALLEY, B. W., and STASTNY, M. (1970). Estrogenic induction of ornithine decarboxylase *in vivo* and *in vitro*. *Science* 170, 336–338.

FAREBROTHER, D. A., and MANN, C. B. (1970). The histological effects of epithelial growth factor and its antiserum in the neonatal rat. *Biochem. J.* 118, 33P.

GREENE, L. A., SHOOTER, E. M., and VARON, S. (1969). Subunit interaction and enzymatic activity of mouse 7S nerve growth factor. *Biochemistry* 8, 3735–3741.

HEIMBERG, M., WEINSTEIN, I., LEQUIRE, V. S., and COHEN, S. (1965). The induction of fatty liver in neonatal animals by a purified protein (EGF) from mouse submaxillary gland. *Life Sci.* 4, 1625–1633.

HERBST, E. J., and BACHRACH, U. (1970). Metabolism and biological functions of polyamines. *Ann. N.Y. Acad. Sci.* 171, 691–1009.

HOOBER, J. K., and COHEN, S. (1967a). Epidermal growth factor. I. The stimulation of protein and nucleic acid synthesis in chick embryo epidermis. *Biochim. Biophys. Acta* **138**, 347–356.

HOOBER, J. K., and COHEN, S. (1967b). Epidermal growth factor. II. Increased activity of ribosomes from chick embryo epidermis for cell-free protein synthesis. *Biochim. Biophys. Acta* **138**, 357–368.

HURWITZ, C., and ROSANO, C. L. (1967). The intracellular concentration of bound and unbound magnesium ions in *Escherichia coli. J. Biol. Chem.* **242**, 3719–3722.

JÄNNE, J., and RAINA, A. (1969). On the stimulation of ornithine decarboxylase and RNA polymerase activity in rat liver after treatment with growth hormone. *Biochim. Biophys. Acta* **174**, 769–772.

JÄNNE, J., RAINA, A., and SIIMES, M. (1968). Mechanism of stimulation of polyamine synthesis by growth hormone in rat liver. *Biochim. Biophys. Acta* **166**, 419–426.

LEVI-MONTALCINI, R., and ANGELETTI, P. U. (1964). Hormonal control of the NGF content in the submaxillary salivary glands of mouse. *In* "Salivary Glands and Their Secretions" (L. M. Sreebny and J. Meyer, eds.), pp. 129–141. Macmillan, New York.

MANN, C. B., and FENTON, E. L. (1970). Biological effects of epithelial growth factor and its antiserum in neonatal rats. *Biochem. J.* **118**, 33P.

OLIVER, W. J., and GROSS, F. (1967). Effect of testosterone and duct ligation on submaxillary renin-like principle. *Amer. J. Physiol.* **213**, 341–346.

PEGG, A. E., LOCKWOOD, D. H., and WILLIAMS-ASHMAN, H. G. (1970). Concentrations of putrescine and polyamines and their enzymic synthesis during androgen-induced prostatic growth. *Biochem. J.* **117**, 17–31.

REYNOLDS, V. H., BOEHM, F. H., and COHEN, S. (1965). Enhancement of chemical carcinogenesis by an epidermal growth factor. *Surg. Forum* **16**, 108–109.

RUSSELL, D., and SNYDER, S. H. (1968). Amine synthesis in rapidly growing tissues: Ornithine decarboxylase activity in regenerating rat liver, chick embryo, and various tumors. *Proc. Nat. Acad. Sci. U.S.* **60**, 1420–1427.

SCHACHTER, M. (1969). Kallikreins and kinins. *Physiol. Rev.* **39**, 509–547.

SREEBNY, L. M., and MEYER, J., eds. (1964). "Salivary Glands and Their Secretions." Macmillan, New York.

STASTNY, M., and COHEN, S. (1970). Epidermal growth factor. IV. The induction of ornithine decarboxylase. *Biochim. Biophys. Acta* **204**, 578–589.

TAYLOR, J. M. (1970). Epidermal growth factor: High and low molecular weight forms. Doctoral Dissertation, Department of Biochemistry, Vanderbilt University, Nashville, Tennessee.

TAYLOR, J. M., COHEN, S., and MITCHELL, W. M. (1970). Epidermal growth factor: High and low molecular weight forms. *Proc. Nat. Acad. Sci. U.S.* **67**, 164–171.

TURKINGTON, R. W. (1969a). The role of epithelial growth factor in mammary gland development in vitro. *Exp. Cell Res.* **57**, 79–85.

TURKINGTON, R. W. (1969b). Stimulation of mammary carcinoma cell proliferation by epithelial growth factor in vitro. *Cancer Res.* **29**, 1457–1458.

TURKINGTON, R. W., MALES, J. L., and COHEN, S. (1971). Synthesis and storage of epithelial-epidermal growth factor in submaxillary gland. *Cancer Res.* **31**, 253–256.

VARON, S., NOMURA, J., and SHOOTER, E. M. (1967). The isolation of the mouse nerve growth factor protein in a high molecular weight form. *Biochemistry* **6**, 2202–2209.

Tumor Angiogensis: Role in Regulation of Tumor Growth[1]

Judah Folkman

Department of Surgery, Children's Hospital Medical Center and
The Harvard Medical School, Boston, Massachusetts

I. INTRODUCTION

The growth of solid neoplasms is always accompanied by neovascularization. Grafts of nonmalignant tissues, such as thyroid or endometrium, do not elicit capillaries from the host (Ehrmann and Knoth, 1968; Merwin and Algire, 1956). Instead, capillary sprouts from the graft join host vessels. Nonmalignant cells may form small nodules when injected into a host animal, but they neither can elicit new host capillary proliferation nor send out capillaries to connect to the vascular system of the host (Giovanella *et al.*, 1972). The behavior of transplanted embryonic tissues is less clear. However, recent evidence suggests that their rapid vascularization is due to the growth of capillary sprouts from the embryonic graft and is not the result of stimulation of neovascularization in the host (Andrews, 1972). Only neoplastic cells, then, seem to have the capacity to elicit from the host new capillaries, vigorously and continuously throughout the life of the tumor. Evidence has recently accumulated to

[1] Supported by Public Health Service Grant 5R01 CA 08185–06 from the National Cancer Institute, by Grant IC-28 from the American Cancer Society, and by gifts from the Merck Company and Alza Corporation.

suggest that this property, *tumor angiogenesis*, may play a critical role in the regulation of growth of solid tumors.

II. INHIBITION OF TUMOR GROWTH IN THE ABSENCE OF ANGIOGENESIS

We have previously reported results of the study of tumor implants in isolated perfused organs (Folkman *et al.*, 1966; Folkman, 1970). Thyroid glands from dogs or rabbits were perfused through their arterial tree with tissue culture serum solutions to maintain the glands in isolated organ chambers for 1–2 weeks. Tumor cells from mice and rats when implanted into these isolated organs developed into tiny nodules that stopped growing at diameters of 1–2 mm. They remained viable but dormant. Arrest of growth was not immunologic. The perfusate contained no cells, and histologic sections failed to show immune rejection.

When these tumors were reimplanted into the host animal from which they came, they grew rapidly to 10–20 times the diameter observed in the isolated organs. The difference between the two growth patterns was *vascularization*. Tumors in isolated organs never became vascularized. Tumors *in vivo* became vascularized after they reached approximately 2 mm in diameter, and then grew rapidly to several centimeters in diameter. The lack of vascularization in the isolated organ was an artifact of the perfusion itself. Capillary endothelial cells in isolated organs began to degenerate during the first day of perfusion (Gimbrone *et al.*, 1969). No mitoses were seen in the capillary endothelium throughout the perfusion. A wound made in the gland did not stimulate capillary proliferation.

It was this artifact of organ perfusion which first made us appreciate that absence of angiogenesis could limit growth of solid tumors to a few millimeters in diameter. Although the experimental system was crude, it provided the first evidence for this concept. Prior to this experiment, there had been no experimental model from which one could conclude that blockade of angiogenesis would lead to inhibition of tumor growth. Three other lines of recent evidence support this hypothesis.

1. Avascular carcinomas were created by anterior chamber implantation in a susceptible rabbit strain (Gimbrone *et al.*, 1972). Unable to elicit new capillary ingrowth, these small spherical tumors suspended in the aqueous humor survived by simple diffusion. Since the aqueous humor continually renews itself, the tumor implants were provided with sufficient nutrients. After a brief period, each implant ceased growing at a diame-

ter of less than 1 mm and then remained dormant indefinitely. By dormancy, we refer to a population of cells in which there is a cell loss ratio of one, that is to say, in which birth and death rates are equal and total population constant. Cells within these implants were viable as demonstrated by (a) DNA synthesis and mitotic figures in the periphery of the mass and, (b) ability of the arrested tumor to grow rapidly when transplanted to a site where vascularization could occur. Whenever a dormant avascular tumor was placed in contact with the iris, vascularization occurred within 36 hours and was followed by tumor growth to several thousand times initial volume with 10–14 days. Volume doubling times in the vascular phase averaged 11.5 hours, compared with 4.95 days in the prevascular phase prior to dormancy.

2. Three tumor lines were grown as spheroids in soft agar (Folkman and Hochberg, 1973). These were a mouse melanoma, a murine leukemia, and a Chinese hamster V-79 cell line. These spheroids were started from a single cell in a flask of soft agar. Every other day each spheroid was changed to new medium and its diameter was measured. All these spheroids entered a dormant phase of a few millimeters in diameter beyond which there was no further growth. The dormant phase continued indefinitely no matter how many times the medium was changed. Histologic sections revealed several layers of intact tumor cells on the periphery containing mitotic figures and [³H]thymidine labeled nuclei. There was a central core of necrosis suggesting that a balance of cell renewal and cell death was reached in these dormant tumors. This histologic pattern was very similar to that of the avascular tumors in the anterior chamber of the eye and also the avascular tumors in the isolated perfused organs.

3. In a recent study from my laboratory (D. Knighton and J. Folkman, unpublished data, 1973), we observed that tumors implanted on the chorioallantoic membrane or the yolk sac membrane of the chick embryo have a relatively prolonged period of avascular growth, depending upon the age of the embryo at the time of tumor implant. Tumor implants made during this refractory period remain dormant at approximately 1 mm in diameter. This was demonstrated for mouse melanoma (B16), rat Walker 256 carcinosarcoma, and the rabbit Brown-Pearce carcinoma. If the original tumor implant was less than 1 mm, it grew up to this size and remained avascular and dormant. If a large tumor was implanted it regressed to 1 mm where it remained dormant but viable. Once capillaries began to proliferate and penetrate the tumor, rapid growth began about 24 hours later. Within five to six days, most tumors attained diameters of 10 mm or more.

These four experimental models indicate that solid tumors growing as three dimensional spheroids or elliptoids reach a diameter of a few millimeters at which the ratio of surface to volume is insufficient to provide adequate uptake of nutrients and release of catabolites. This surface: volume ratio becomes limiting to further growth regardless of how much exchange of nutrient or catabolite goes on near the surface of the spheroid or how closely this surface is apposed to a vascular network. New capillaries must *penetrate* the tumor nodule before such a spheroid can overcome its diffusion limitations. By whatever means angiogenesis is prevented, we have proposed the term "antiangiogenesis" to epitomize this concept (Folkman, 1972). It is conceivable that, if tumor angiogenesis *in vivo* could be interrupted under any condition, not just in the special situation of the anterior chamber of the eye, many types of solid tumors might be kept dormant at a tiny diameter. Such an approach would require a detailed understanding of the mechanism of tumor angiogenesis. What is known about this mechanism is outlined below.

III. STIMULATION OF CAPILLARY PROLIFERATION BY TUMOR OCCURS AT A DISTANCE

The literature on tumor angiogenesis is extensive, but mostly descriptive (Algire, 1943; Goodall *et al.*, 1965; Warren, 1966, 1968; Feigin *et al.*, 1958). Tumor implants put into chambers in the rabbit ear or the hamster cheek pouch elicit capillary proliferation. Greenblatt and Shubik (1968) were the first to suggest that tumor angiogenesis might be mediated by a diffusible factor which could traverse a Millipore filter with a pore size of 0.4 μm. Ehrmann and Knoth (1968) later confirmed this. We have had similar experience with tumors contained in Millipore chambers implanted into the subcutaneous dorsal air pouch in the rat. New capillaries appear beneath the Millipore filter, presumably in response to a factor diffusing through the filter.

Recently we have shown that tumor stimulation of capillary proliferation occurs across distances greater than the thickness of a standard Millipore filter (Cavallo *et al.*, 1972). When minute volumes (0.03 ml) of tumor cells were injected in the subcutaneous air sac in the rat, DNA synthesis began in previously resting endothelial cells, as early as 6–8 hours later, in vessels 1–3 mm from the site of implantation. Electron microscopic autoradiography showed incorporation of [³H]thymidine in endothelial cells surrounding the tumor implant. No tumor cells were present in any section, nor were there inflammatory cells. Many sections were taken at distances of at least 3 mm from the nearest tumor cell.

By 48 hours, new blood vessel formation was detected next to the living tumor. Dead tumor cells did not stimulate growth of capillary endothelium.

In another study (Gimbrone *et al.*, 1973a), tumor implants made in the cornea produced neovascularization in the iris of the rabbit eye. The maximum distance of tumor from the vascular bed was 5 mm. When tiny implants are made in the rabbit cornea, separation of tumor cells from the responding vascular bed is assured because there are no direct vascular of lymphatic connections between the corneal stroma and the iris stroma. Therefore, any proposed diffusible material factor from the tumor must have diffused through the corneal and iris stromas and/or via the slowly circulating aqueous humor to reach the responding vascular bed.

IV. EVIDENCE FOR HUMORAL TRANSMISSION OF TUMOR ANGIOGENESIS: TUMOR ANGIOGENESIS FACTOR

For the above evidence that tumor can stimulate capillary endothelial mitosis at a distance, we considered two possible explanations. (1) Tumor cells release a diffusible factor which is mitogenic to capillary endothelial cells directly or through some intermediate pathway; or, (2) tumor cells consume some ubiquitous "inhibitor" of capillary proliferation, thus producing a local zone deficient in the inhibitor, which releases capillary proliferation. The latter hypothesis, i.e., that tumor implants act as a "sink" for an inhibitor of capillary proliferation, has little evidence to support it and at least one argument against it. For a tumor implant in the cornea to stimulate iris neovascularization by this route, it would be necessary to say that the tiny tumor had consumed all of the so-called inhibitor in the aqueous humor in spite of the constant renewal of this humor. Nevertheless, a decisive experiment to rule out this "negative feedback hypothesis" has not been done.

On the other hand, we do have increasing evidence that tumors release a diffusible factor which itself is mitogenic to capillary endothelium. We have previously shown that both the cytoplasm and nuclei of the tumor cells from animal and man, have the capacity to stimulate endothelial mitosis during capillary proliferation in the absence of intact tumor cells (Folkman *et al.*, 1971). Walker 256 carcinoma cells removed either from the ascites or solid form of the tumor were washed and then lysed by nitrogen cavitation. The nuclei were separated, and the remaining cellular components were sedimented by centrifugation at 260,000 g. Both the nuclear component and the cytoplasmic component had angiogenesis activity when tested in the subcutaneous dorsal air sac of the rat. Crude purification of the cytoplasmic factor was accomplished by delipidation, trypsinization, and then gel filtration on Sephadex G-100. Fraction II (Folkman *et al.*, 1971) of this eluate contained angiogenesis activity. However, purification beyond this was difficult. When further chromatographic separation was attempted, angiogenesis activity spread over several fractions as though the tumor angiogenesis factor (TAF) were attached to some carrier protein. We have not yet overcome this obstacle.

However, significant progress has been made in purification of TAF in the nuclear component. A procedure has been developed (Tuan *et al.*, 1973) in which purified chromatin from tumor nuclei was separated into DNA and chromatin proteins on a Bio-gel A 5-m column. Angiogenesis activity was associated with the chromatin, but not with the separated DNA. The chromatin was further fractionated on a CM-Sephadex column into histone and nonhistone proteins. The nonhistone proteins demonstrated potent angiogenesis activity. This activity is contained in a pool of approximately 20 protein moieties as determined by acrylamide gel electrophoresis. RNA is present at 3 parts per 100. Incubation with ribonuclease inactivates activity. The histone proteins show no angiogenesis activity. The macromolecular composition and amino acid analysis of both the histone and nonhistone proteins have been determined. Nonhistone proteins from normal tissue, such as rat liver, have no activity. Likewise, histone proteins from liver are also negative. Studies are now under way to purify the pool of nonhistone proteins in an attempt to locate angiogenesis activity to a single protein moiety.

V. BIOASSAYS FOR TUMOR ANGIOGENESIS FACTOR

A. Subcutaneous Dorsal Pouch

TAF in the cytoplasm was originally assayed in the subcutaneous dorsal air sac of the rat (Folkman *et al.*, 1971). The subcutaneous space was inflated with 30 cc of air. A tiny silicone tube was implanted into this space and sutured to the fascial floor of the sac. Two milliliters of the fraction to be tested were pumped with a syringe intermittently into the sac over a 48-hour period. The rats were restrained throughout this time. The restraint was necessary to prevent rats from biting the tubing. Later, it was found that restraint also induced endogenous release of corticosteroids and prevented background inflammation. Only tumors or TAF fractions were capable of eliciting capillary proliferation in restrained rats or in unrestrained rats treated with high doses of corticosteroid. This assay was cumbersome, necessitated strict sterile technique, and required large amounts of test material (500 μg of total protein). It was reliable in that false-positive results from nonspecific inflammation were not observed. However, this assay became less useful as purification of nuclear TAF proceeded. Despite the great potency of these fractions, the rat back assay did not detect concentrations of activity in the range of 10–20 µg of total protein. Therefore, a more sensitive assay was recently developed in which the test fraction is inserted into the cornea of the rabbit eye.

B. Corneal Pocket

A small pocket 1 mm × 2 mm was made in the cornea of the rabbit eye with a spatula (Gimbrone et al., 1973a). The blind (Fig. 1) end of the pocket was placed 2.0 mm from the edge of the cornea (limbus). The pocket was filled with 0.03 ml of 5% acrylamide containing the test fraction. This acrylamide was buffered to 7.5 with HEPES as previously described (Folkman et al., 1972) except that Ringer's solution was the solvent, the final concentration after polymerization was 5.0%, and all

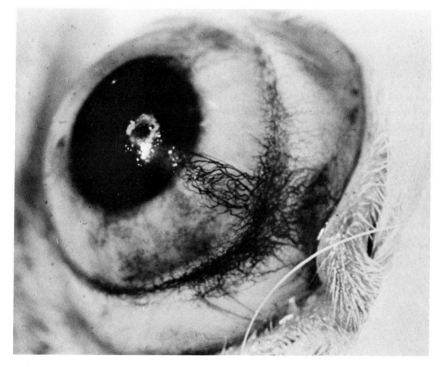

Fig. 1. New capillaries in cornea of rabbit eye. The capillaries have grown from the limbal edge of cornea toward pellet of acrylamide implanted 20 days previously. Of the nonhistone protein containing TAF (Walker 256 carcinosarcoma), 0.5 µg were dispersed in the soft acrylamide. The volume of acrylamide was 20 µl, and it was implanted 2.5 mm from the edge of the cornea.

reagents were sterilized by Millipore filtration. One volume of the fraction to be tested was added to four volumes of the liquid acrylamide, and this was allowed to polymerize by the addition of ammonium persulfate and sodium bisulfite. The polymerized mass had the consistency of clear paste and could be squeezed through a No. 25 hypodermic needle into the corneal pocket. The acrylamide polymer is absolutely inert both *in vivo* and in tissue culture. It provided a slow release depot for the test proteins over a period of 10–15 days. When 0.5–1.0 μg of the non-histone protein was inserted into the corneal pocket, new capillaries began to penetrate the cornea by day 3 and continued to grow at 1 mm per day for approximately 20 days. The new capillaries were observed every other day with a horizontal stereoscope slit lamp. Histologic sections showed these to be proliferating capillaries with endothelial cells in mitosis. There was no inflammation. Histone proteins placed in acrylamide in the opposite eye showed no effect throughout this period. This assay is very precise and reliable. Also the difference between pure angiogenesis and the type of vasoproliferation secondary to inflammation can be quickly determined in the eye. Inflammation, for example (formic acid in the acrylamide) causes edema, loss of corneal transluscency, and the capillaries grow randomly over a wide area surrounded by heavy white exudate. On the other hand, pure angiogenesis in the absence of inflammation, demonstrates rapidly advancing capillary loops and "hair pins" in a clear, transluscent cornea without exudate.

C. Endothelial Cells in Culture

We do not yet have an *in vitro* assay for TAF. We have developed cultures of endothelial cells toward this goal. Segments of aorta from the rat have been cultured (Sade *et al.*, 1972). Endothelial cells were capable of incorporating [³H]thymidine as late as 6 days after beginning of culture. However, trauma during removal of the aortas from rats also stimulated DNA synthesis in endothelial cells. This caused a high background level of [³H]thymidine labeling, and made it difficult to test fractions for their ability to stimulate increased DNA synthesis.

More recently, Gimbrone *et al.* (1973b) have maintained human endothelial cells in culture by modifying a method first developed by R. Nachman (Jaffee *et al.*, 1972). Umbilical cords were obtained fresh from the delivery room. The umbilical vein was flushed with Ringer's solution, filled with collagenase (0.2%), and incubated for 30 minutes at 37°C. Endothelial cells were harvested as a pellet by flushing the vein with Hank's salt solution (free of calcium and magnesium). Approximately 5×10^6 cells were obtained in this way. They were cultured in Medium 199 with 30% fetal calf serum, in multiple wells of the Falcon microtray. These have been proved to be endothelial cells by electron microscopy and by staining with fluorescent antithrombasthenin antisera. They have been carried through 12 subcultures. When confluent, the labeling index with [³H]thymidine was approximately 6%. Nonhistone protein pool containing TAF starting at 0.1 µg/ml was added to confluent cultures. The cultures were exposed to [³H]thymidine after 48 hours of TAF and then autoradiographs made. The ratio of labeled to unlabeled cells was determined. Results so far are too preliminary to report, but the method of culture is now reliable enough so that it may possibly serve as an *in vitro* assay for any endothelial cell mitogen. In the final analysis, however, any purified fraction must be retested *in vivo*, preferably in the cornea; it must be capable of inducing capillary proliferation.

VI. SUMMARY

This study indicates that solid tumors appear to release a diffusible factor, tumor angiogenesis factor (TAF) which is mitogenic to endothelial cells and can induce capillary proliferation at distances of up to 3–5 mm from the nearest tumor cell. Normal tissues are incapable of this. Once new capillaries have penetrated a solid tumor implant, the tumor grows rapidly and continuously. In the absence of penetration by new capillaries, i.e., in the absence of angiogenesis, most solid tumors appear to enter a dormant phase indefinitely and stop expansion at a diameter of a few millimeters. We have proposed that, if TAF could be inhibited, most solid tumors might remain dormant. This concept of "antiangiogenesis" has been derived from our increased understanding of the mechanism of tumor growth prior to vascularization.

REFERENCES

ALGIRE, G. H. (1943). Microscopic studies of the early growth of a transplantable melanoma of the mouse, using the transparent-chamber technique. *J. Nat. Cancer Inst.* **4,** 13–20.

ANDREWS, E. J. (1972). The survival and differentiation of embryonic lung tissues in diffusion chambers and in mammary fat pads of mice. *Anat. Rec.* **172,** 89–96.

CAVALLO, T., SADE, R., FOLKMAN, J., and COTRAN, R. S. (1972). Tumor angiogenesis: Rapid induction of endothelial mitoses demonstrated by autoradiography. *J. Cell Biol.* **54,** 408–420.

EHRMANN, R. L., and KNOTH, M. (1968). Choriocarcinoma: Transfilter stimulation of vasoproliferation in the hamster cheek pouch—studied by light and electron microscopy. *J. Nat. Cancer Inst.* **41,** 1329.

FEIGIN, I., ALLEN, L. B., and LIPKIN, O. (1958). The endothelial hyperplasia of the cerebral blood vessels with brain tumors and its sarcomatous transformation. *Cancer* 11, 264.

FOLKMAN, J. (1970). The intestine as an organ culture. *In* "Carcinoma of the Colon and Antecedent Epithelium" (W. J. Burdette, ed.), Chapter 6, pp. 113–127. Thomas, Springfield, Illinois.

FOLKMAN, J. (1972). Anti-angiogenesis: New concept for therapy of solid tumors. *Ann. Surg.* 175, 409–416.

FOLKMAN, J., and HOCHBERG, M. (1973). Self-regulation of growth in three dimensions. *J. Exper. Med.* 138, 745–753.

FOLKMAN, J., COLE, P., and ZIMMERMAN, S. (1966). Tumor behavior in isolated perfused organs: In vitro growth and metastases of biopsy material in rabbit thyroid and canine intestinal segment. *Ann. Surg.* 164, 491–502.

FOLKMAN, J., MERLER, E., ABERNATHY, C., and WILLIAMS, G. (1971). Isolation of a tumor factor responsible for angiogenesis. *J. Exp. Med.* 133, 275–288.

FOLKMAN, J., CONN, H., and HARMEL, R. (1972). Acrylamide polymerization: New method for determining the oxygen content of blood. *Science* 178, 170–172.

GIMBRONE, M. A., JR., ASTER, R. H., COTRAN, R. S., CORKERY, J., JANDL J. H. and FOLKMAN, J. (1969). Preservation of vascular integrity in organs perfused in vitro with a platelet-rich medium. *Nature (London)* 222, 33–36.

GIMBRONE, M. A., JR., LEAPMAN, S., COTRAN, R. S., and FOLKMAN, J. (1972). Tumor dormancy in vivo by prevention of neovascularization. *J. Exp. Med.* 136, 261–276.

GIMBRONE, M. A., JR., LEAPMAN, S. B., COTRAN, R. S., and FOLKMAN, J. (1973a). Tumor angiogenesis: Iris neovascularization at a distance from experimental intraocular tumors. *J. Nat. Cancer Inst.* 50, 219–228.

GIMBRONE, M. A., JR., COTRAN, R. S., and FOLKMAN, J. (1973b). Endothelial regeneration and turnover. Studies with human endothelial cell cultures. *Microvascular Res.* 6, 249 (Abstract).

GIOVANELLA, B. C., YIM, S. O., STEHLIN, J. S., and WILLIAMS, L. J., JR. (1972). Development of invasive tumors in the "nude" mouse after injection of cultured human melanoma cells. *J. Nat. Cancer Inst.* 48, 1531–1533.

GOODALL, C. M., FELDMAN, R., SANDERS, A. G., and SHUBIK, P. (1965). Vascular patterns of four transplantable tumors in the hamster (*Mesocricetus auratus*). *Angiology* 16, 622–625.

GREENBLATT, M., and SHUBIK, P. (1968). Tumor angiogenesis: Transfilter diffusion studies in the hamster by transparent chamber technique. *J. Nat. Cancer Inst.* 41, 111.

JAFFE, E. A., NACHMAN, R. L., BECKER, C. G., and MINICK, R. C. (1972). Culture of human endothelial cells derived from human umbilical cord veins. *Circulation* 46, 252 (Abstract).

MERWIN, R. M., and ALGIRE, G. H. (1956). The role of graft and host vessels in the vascularization of grafts of normal and neoplastic tissue. *J. Nat. Cancer Inst.* 17, 23–33.

SADE, R. M., FOLKMAN, J., and COTRAN, R. S. (1972). DNA synthesis in endothelium of aortic segments in vitro. *Exp. Cell Res.* 74, 297–306.

TUAN, D. T., MERLER, E., SMITH, S., and FOLKMAN, J. (1973). Isolation of the nonhistone proteins of rat Walker carcinoma 256: Their association with tumor angiogenesis. *Biochemistry* 12, 3159–3165.

WARREN, B. A. (1966). The ultrastructure of capillary sprouts induced by melanoma transplants in the golden hamster. *J. Roy. Microsc. Soc.* [3] 86, 177–187.

WARREN, B. A. (1968). In vivo and electron microscopic study of vessels in a haemangiopericytoma of the hamster. *Antiologica* 5, 230–249.

Hormonal Control of Oviduct Growth and Differentiation

B. W. O'Malley,[1] A. R. Means,[1] S. H. Socher,[1] T. C. Spelsberg,[2]
F. Chytil,[3] J. P. Comstock,[1] and W. M. Mitchell[4]

*Departments of Obstetrics and Gynecology, Biochemistry, Medicine,
Physiology, and Microbiology, Vanderbilt University School of Medicine,
Nashville, Tennessee*

I. INTRODUCTION

The immature chick oviduct has proved to be an excellent model system for studies of the mechanisms of the female sex steroids estrogen and progesterone (O'Malley *et al.*, 1969; O'Malley, 1967, 1969). This tissue can be used successfully to investigate cytodifferentiation which is not limited to the period of embryonic development. Under the influence of estrogen the immature oviduct begins to differentiate morphologically and biochemically. The synthesis of tissue-specific nucleic acids and proteins is stimulated in response to this steroid. In contrast, progesterone

[1] Present address: Department of Cell Biology, Baylor University College of Medicine, Houston, Texas.

[2] Present address: Department of Endocrine Research, Mayo Clinic, Rochester, Minnesota 55901.

[3] Present address: Department of Biochemistry, Vanderbilt University School of Medicine, Nashville, Tennessee 37203.

[4] Present address: Departments of Microbiology and Medicine, Vanderbilt University School of Medicine, Nashville, Tennessee 37203.

administration to estrogen-pretreated animals induces synthesis of the specific oviduct protein, avidin.

II. ESTROGEN-MEDIATED DIFFERENTIATION

When the synthetic estrogen, diethylstilbestrol (DES), or natural estrogens are injected daily into immature chicks, three new cell types can be identified in the oviduct mucosa (O'Malley *et al.*, 1969; Kohler *et al.*, 1969). (1) Tubular glands which secrete ovalbumin and lysozyme begin to appear between days 2 and 4 of treatment. (2) At 6 days of hormone administration, ciliated cells can be found. (3) It is not until 7–9 days of hormone treatment that the goblet cells, which produce avidin, appear. Avidin, however, is not detected at any stage of development until a single dose of progesterone is given. Estrogen is thus responsible for the morphological differentiation of new cell types, cellular hypertrophy, and also for the induction of synthesis of cell-specific proteins. How this simple phenolic steroid can accomplish this feat has been the topic of much of our research.

As has been found in other steroid hormone target tissues, the oviduct contains an estrogen-binding macromolecule which is tissue specific, heat labile, and binds 17β-estradiol or DES with an affinity ($K_d \simeq 10^{-10}$) greater than estrone and estriol (W. L. Eaton and B. W. O'Malley, unpublished data). This receptor protein has little or no affinity for androgens, progesterone, or glucocorticoids. It is presently our opinion that all steroid hormones most likely act in target cells by first combining with a cytoplasmic receptor, followed by transfer of the hormone–receptor complex into the nucleus, where it exerts its primary influence.

Since estrogen is responsible for growth of the oviduct, an early effect on certain key enzymes and structural proteins would be expected. Indeed, a stimulation of the activity of oviduct ornithine decarboxylase, completely blocked by cycloheximide, has been found within hours after *in vivo* or *in vitro* estrogen administration (Cohen *et al.*, 1970). Stimulation of this enzyme and polyamine synthesis have also been linked with the chemical induction of growth in other cell types, and it is possible that polyamines may trigger, or more probably simply support, the necessary cell processes which lead to rapid nucleic acid and protein synthesis and subsequent cell growth.

In addition to this early stimulation of ornithine decarboxylase activity and the later induction of the cell-specific proteins, lysozyme and ovalbumin, estrogen is known to stimulate overall protein synthesis in the oviduct (O'Malley and McGuire, 1968a,b). Since proteins are synthesized on polyribosomes and require tRNA, we have characterized both of these particles with respect to quantity and functional capacity during estrogen-mediated differentiation of the oviduct (Means and O'Malley, 1971; O'Malley et al., 1968). For the first 7 days of hormone treatment, there is an increase in the oviduct content of ribosomes followed by a decline by day 10, when the rate of cytodifferentiation is beginning to plateau. Similarly, a striking increase of tRNA is noted during this same period (O'Malley et al., 1968).

The polysome profiles from the unstimulated chick oviduct also reveal an effect of estrogen, which could account for the stimulation of cell protein synthesis (Means et al., 1971). In the unstimulated oviduct, monomers are the predominant form, but from oviducts treated for 4 days with hormone, the polysome profile reveals a large portion of aggregates. Withdrawal of estrogen at any point of differentiation results in polysomal disaggregation, cessation of protein synthesis, and the growth process is brought to an abrupt halt.

By another approach, we have shown that polyribosome preparations synthesize peptides in a cell-free incorporation system at a maximum level on the fourth day of hormone treatment. By day 7 there is a decline in protein-synthesizing ability (Means et al., 1971). Analysis by polyacrylamide gel electrophoresis of peptides synthesized in vivo or in vitro shows that striking changes occur in the peptides synthesized before and after estrogen treatment (Means et al., 1971). Since the effects are apparent in the in vitro system, it would seem that the hormone must be responsible in some way for affecting the messenger RNA population of the cells of the oviduct mucosa. Indeed, we have found that elements of the transcriptional apparatus are affected prior to the first observance of stimulation of protein synthesis found 3–4 days after hormone treatment has begun.

A very early stimulation of nuclear RNA polymerase activity occurs after a single dose of estrogen and reaches a peak between 12 and 24 hours after treatment (McGuire and O'Malley, 1968). There is an increase in total template capacity to synthesize RNA in vitro, and, by nearest neighbor frequency analysis, we have shown that there is a qualitative change in the RNA product synthesized from the DNA-chromatin template at various stages of estrogen-induced differentiation (O'Malley and McGuire, 1968a; O'Malley et al., 1969).

The next step in the sequence of experiments to determine whether indeed estrogen is active at the level of transcription would be to show that the hormone directs the synthesis of new specific messenger RNA's (mRNA). However, techniques are not available at the present time to accomplish this and so, as an alternative approach, we have examined the translational activity of total mRNA during hormone-induced growth and differentiation of the oviduct (Means *et al.*, 1971). Using a cell-free amino acid incorporation system from a bacterial source and oviduct nuclear RNA, we have found that this RNA exhibits increased messenger activity after a single injection of DES; maximum activity occurs on the third day of treatment. Again, this effect precedes those of the hormone on polysomal protein synthesis and points to an initial effect of estrogen on gene transcription. If this is the case, new species of RNA not present prior to estrogen treatment should be found in the hormone-stimulated oviduct. Using competition analysis in DNA–RNA hybridization experiments we have confirmed that new species of RNA (repeating sequences) do appear with estrogen treatment (O'Malley and McGuire, 1968a). Thus both quantitative and qualitative alterations in gene transcription occur after administration of estrogenic steroid hormones and prior to morphological or biochemical differentiation (O'Malley, 1969).

III. THE CELL CYCLE

Since the chick oviduct approximates a model experimental system for analysis of hormone effects on cell proliferation during development of target tissues, the early effects of estrogen and progesterone on cell division have been examined. The studies reported here are restricted to characterizing the mitotic behavior of the surface epithelium.

A. *Immature (Unstimulated) Oviduct*

Mitotic Activity. A necessary prerequisite for the study of hormone effects on mitosis in a target tissue is a knowledge of the level and chronology of mitotic activity and cell cycle parameters in the absence of hormonal influence. In the immature oviduct, mitoses are infrequent (Kohler *et al.*, 1969). In the present studies the mean mitotic index (MI) was 0.43 in 7-day-old chicks. This reflects the slow natural growth of the oviduct that continues until sexual maturation. The MI remained constant over a 24-hour period. Thus changes observed after hormone treatment should be due to the hormones, not to any natural mitotic rhythms within the tissue.

G_2 *Phase.* DNA synthetic inhibitors, 5-fluorodeoxyuridine (FUdR) and hydroxyurea (HU), have been used to estimate the duration of G_2 in the unstimulated oviduct. Since these antimetabolites block cells in S, the only cells that enter mitosis after treatment should be G_2 cells. FUdR and HU both produce a drop in MI (Fig. 1). The time interval between injection of the inhibitor and the time at which the MI drops

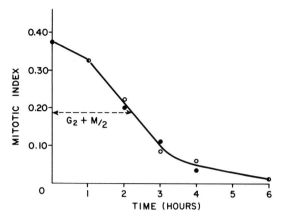

FIG. 1. Mitotic indexes in the surface epithelium after treatment with 5-fluoro-deoxyuridine (●), 15 mg/chick or (○), hydroxyurea 150 mg/chick. Socher and O'Malley (1973).

to 50% of the control value provides an estimate of the mean duration of G_2 + (mitosis/2) (Socher and Davidson, 1971). In the present experiments the duration of G_2 + (mitosis/2) = 2.25 hours (Fig. 1). If it is assumed that the duration of mitosis is 1 hour, then G_2 = 1.75 hours. Another group of chicks received a single injection of [³H]TdR. Labeled mitoses were found 1.5 and 2 hours after administration of the pulse label. Both inhibitor treatments and [³H]TdR experiments indicate that G_2 is short in the immature oviduct, lasting 1.5–2 hours.

S Phase. Unstimulated chicks were given a single injection of [³H]TdR to determine the number of cells in S, i.e., synthesizing DNA. Labeling indexes (LI) were determined 0.5 to 2 hours after the pulse. Grain intensity of labeled nuclei was similar at all times examined. The mean LI of surface epithelium was 3.16, indicating that a small fraction of the cells are synthesizing DNA.

The duration of the S phase can be estimated if the following conditions apply to the population of cells making up the epithelium: (a) The dividing cells constitute a single population of cells, (b) The duration of cell cycle and of its subphases are constant, and (c) The proportion of dividing cells is constant. If these conditions apply to the epithelium, the following relationships hold: (1) MI = d_m/C, (2) LI = d_s/C and therefore (3) $d_m/MI = d_s/LI$, where MI = mitotic index, LI = labeling index, d_m = duration of mitosis, d_s = duration of S, and C = mean cycle duration. Solving Eq. 3 for d_s (assuming d_m = 1 hour), d_s = 7.3 hours in the unstimulated oviduct.

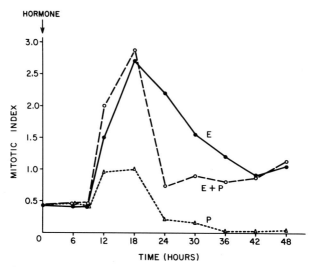

FIG. 2. Mitotic indexes in the surface epithelium following a single injection of 1 mg/chick of either 17β-estradiol (*E*), or progesterone (*P*), or 17β-estradiol and progesterone (*E* + *P*). Socher and O'Malley (1973).

B. Stimulated Oviduct

Mitotic Activity. Immature chicks (7 day) were treated with estrogen and progesterone, alone or in combination, to examine the action of these hormones on cell proliferation. The effects of single and repeated injection of hormones on mitotic activity can be seen in Figs. 2 and 3.

A single injection of estrogen stimulates mitosis in the oviduct (Fig. 2). There is a rapid rise in MI between 9 and 12 hours. The MI reaches a peak at 18 hours after treatment. The level of mitotic activity begins to fall at 24 hours and continues to drop until 42 hours. The frequency of cells in mitosis 48 hours after estrogen is more than twice that observed in the unstimulated oviduct.

When chicks are given a second injection of estrogen 24 hours after the first, a second rise in MI is observed (Fig. 3). The patterns of changes in MI following single and double treatments could be explained in terms of a hormone-induced stimulation of division in a single population of cells. The cells appear to progress through the cycle in a parasynchronous fashion. When chicks receive a single injection of estrogen, the fall in MI may be due to a progression through the cycle at a slower rate, a loss of synchrony or a loss of the ability to divide. However, when chicks are treated twice with estrogen, the cells progress through the cycle at maximal, or nearly maximal, rate and maintain their synchrony.

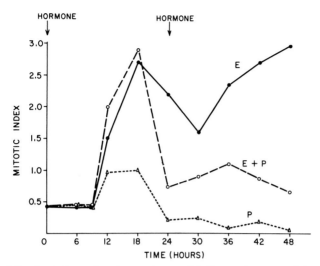

FIG. 3. Mitotic indexes in the surface epithelium after two injections of 1 mg/chick of either 17β-estradiol (E), or progesterone (P), of 17β-estradiol and progesterone (E + P) at 9 and 24 hours. Socher and O'Malley (1973).

Progesterone also stimulates a small fraction of cells to undergo mitosis (Fig. 2). A small, but significant rise in MI is observed 12 and 18 hours after treatment. The frequency of cells in mitosis drops below control level at 24 hours and remains low until 48 hours, with or without a second injection of progesterone (Figs. 2 and 3). These data indicate that progesterone can act as a proliferative stimulus for a population of cells in the surface epithelium of the immature oviduct. Moreover this steroid appears to inhibit the progress through the cycle of normally proliferating cells. The basis and relationship between these two actions of progesterone are not clear.

The initial response to a combination treatment of estrogen and progesterone is similar to that observed with estrogen alone (Fig. 2). However, between 18 and 24 hours after treatment with estrogen and progesterone, a sharp fall in MI is observed. The timing of this drop in mitotic activity is similar to that observed with progesterone alone. With or without a second injection, only small fluctuations in MI are observed 24–48 hours after the beginning of treatment (Figs. 2 and 3). During this time the level of mitotic activity is higher than the control level. It appears that initially estrogen and progesterone in combination act together to stimulate cells to enter mitosis. However, after the initial stimulation, an inhibition of mitosis is observed. This inhibition appears to be due to the action of progesterone.

C. Site of Hormone Stimulation in the Cell Cycle

Within a variety of tissues it has been shown that after an appropriate stimulus, nonproliferating cells can be stimulated to divide. In general, proliferative stimuli act to stimulate G_1 cells to enter S, G_2, and then mitosis, or G_2 cells to undergo mitosis.

To ascertain at which stage the hormone-responsive cells exist in the immature oviduct, experiments were performed in which chicks received simultaneous injections of hormone and either FUdR or HU. If the hormones act to stimulate G_1 cells to enter S, the inhibitors of DNA synthesis should block the hormone-induced stimulation of mitosis. If these hormones stimulate a G_2 population to undergo mitosis, the inhibitors should not block the early hormone induced rise in MI. Both FUdR and HU inhibit the stimulation of mitosis that normally occurs following treatment with estrogen and progesterone, alone or in combination (Fig. 4). This demonstrates that the proliferative stimuli act prior to the completion of DNA synthesis, that is, in G_1 or S. In the unstimulated oviduct $G_2 = 1.75$ hours and S = 7.3 hours. Furthermore a rise in MI is not observed until 12 hours after administration of the hormones. It appears, therefore, the estrogen and progesterone, alone or in combination, stimulate G_1 cells to enter S. This is similar to the action of estrogen as a

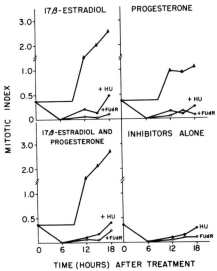

Fig. 4. Effects of 5-fluorodeoxyuridine (FUdR; 15 mg/chick) or hydroxyurea (HU; 150 mg/chick) on changes in mitotic indexes induced by 1 mg/chick of either 17β-estradiol, or progesterone, or both. Socher and O'Malley (1973).

proliferative stimulus in the uterine and vaginal epithelia in the mouse (Perrotta, 1962).

Tubular Gland Formation. Estrogen stimulates the formation of tubular glands which develop as budlike invaginations from the surface epithelium. With estrogen alone, tubular glands begin to form as early as 36 hours after the beginning of treatment, and numerous developing glands are present at 48 hours. No tubular gland formation was observed in oviducts treated with progesterone alone or in combination with estrogen. Progesterone acts as an inhibitor of morphogenetic movements. Oka and Schimke (1969a,b) have described an antagonism of estrogen-induced tubular gland development by progesterone. The results presented here are in agreement with those studies. However, we have shown that progesterone can act initially in combination with estrogen to stimulate proliferation.

Estrogen and progesterone, alone or in combination, stimulate cells to enter mitosis in the immature chick oviduct. However, progesterone inhibits division and morphogenetic movements. All three hormone treatments appear to act in G_1 by stimulating cells to initiate DNA synthesis. The identification of the populations of cells stimulated to divide by these hormones is of importance in the understanding of cytodifferentiation of the surface epithelium of the oviduct.

IV. PROPERTIES OF OVIDUCT CHROMATIN

The newly differentiated tubular gland cells synthesize specific proteins such as ovalbumin and lysozyme in response to estrogen administration. Previous data have suggested transcriptional control of both the differentiation process and of cell-specific protein synthesis. Based upon the hypothesis that changes in gene transcription during differentiation reflect, in part, changes in the tissue-specific pattern of gene restriction, we have investigated changes that occur in the chemical composition, physical properties, and template capacity of oviduct chromatin during estrogen-mediated tissue differentiation. We have separately analyzed changes in the components of chromatin, i.e., DNA, RNA, histone proteins, and non-histone (acidic) protein. The following paragraph presents in abstract form the changes in quantities, immunochemical identity, and configurational relationship to DNA of the nuclear proteins which occur during oviduct development.

Chromatin was isolated from purified nuclei of chick oviduct at various stages of estrogen-mediated development. The chromatin was then dissolved in a high salt–urea solvent at low pH to dissociate the histones and the acidic protein fraction AP_1. After centrifugation the supernatant was dialyzed against H_2O and acidified to precipitate the acidic protein AP_1. The DNA-acidic protein complex, sedimented during centrifugation, was then resuspended in dilute buffer and the insoluble acidic protein (AP_2) removed by low speed centrifugation. The third acidic protein fraction (AP_3) was next isolated by treating the nucleoacidic protein (i.e., $DNA–AP_3–AP_4$ complex) with high salt–urea (buffered at pH 8.5) and centrifuged. The pellet contained the DNA still in association with the fraction AP_4; the supernatant fraction contained fraction AP_3.

Quantitative analysis of the chromatin from various stages of oviduct development demonstrated that while the levels of histones varied randomly, the levels of the total acidic chromatin proteins increased during the first few days of differentiation followed by a gradual decrease until completion of development (15 days of DES treatment) (Table 1). The levels of chromatin-associated RNA followed a similar pattern. Moreover the capacity of the intact chromatins to serve as templates for *in vitro* RNA synthesis using bacterial polymerase also increased during the first few days of differentiation and then decreased during the final stages of development (Table 1). The nucleoacidic protein preparations, in which the fractions AP_3 and AP_4 were complexed to DNA, also displayed an increase in the levels of acidic protein during the first few days of development followed by a gradual decrease during the remaining days of development. Since the amount of AP_4 is extremely small, the major changes in the level of acidic protein which occur in chromatin during development appear to be due largely to changes in the AP_3 fraction (Table 1). Interestingly, the acidic proteins seem to restrict a fraction of the DNA during the early stages of differentiation. This is demonstrable by changes in the template capacity of the various nucleoacidic protein preparations (Table 1). In order to further investigate this restriction the AP_3 fraction was analyzed in more detail using a newly developed immunological approach.

A. Immunochemistry

The lack of tissue specifity of DNA (Gurdon, 1962) and of histones (Hnilica, 1967) suggested that other components of chromatin may be responsible for tissue specific differences in the expression of the information stored in DNA. The question was therefore asked whether the tissue specifity could reside in nonhistone (acidic) proteins (Chytil and Spelsberg, 1971). To clarify the structural relationship and the role the acidic proteins in the tissue specific restriction of DNA, the antigenic properties of nonhistone protein–DNA complexes isolated from different

TABLE 1

CHEMICAL COMPOSITION AND TEMPLATE CAPACITY OF TREATED AND
UNTREATED CHROMATIN PREPARATIONS FROM THE CHICK OVIDUCT
AT VARIOUS STAGES OF DEVELOPMENT[a]

Chromatin type	Days of DES treatment	Histone: DNA	Nonhistone protein: DNA	RNA:DNA	N moles [^{14}C]UMP incorporated per mg DNA
Intact	0	1.06 ± 0.04	0.87 ± 0.01	0.115 ± 0.013	9.1 ± 1.6
chromatin	4	0.99 ± 0.08	1.18 ± 0.03	0.145 ± 0.008	14.9 ± 0.7
	7	1.00 ± 0.09	1.08 ± 0.04	0.180 ± 0.015	14.0 ± 1.4
	12	1.10 ± 0.15	0.82 ± 0.01	0.085 ± 0.010	11.4 ± 1.5
	14	0.94 ± 0.05	0.76 ± 0.04	—	—
	19	1.10 ± 0.16	0.53 ± 0.03	0.081 ± 0.012	11.0 ± 0.2
Nucleo-acidic	0	—	0.33 ± 0.02		1950 ± 20
protein	4	—	0.50 ± 0.04		1950 ± 110
(minus	7	—	0.49 ± 0.08		2270 ± 30
AP_1, AP_2)	12	—	0.43 ± 0.05		2200 ± 50
	19	—	0.43 ± 0.04		2060 ± 10
Deproteinized	0	—	0.022 ± 0.02		2340 ± 60
chromatin	4	—	0.065 ± 0.02		2210 ± 100
(minus	7	—	0.042 ± 0.03		2323 ± 177
AP_1, AP_2,	12	—	0.032 ± 0.03		2345 ± 95
AP_3)	19	—	0.036 ± 0.03		2070 ± 110
Pure DNA (minus AP_1, AP_2, AP_3, AP_4)		—	—		2250 ± 90

[a] Values are expressed as the range of three analyses. Procedures for the analysis of chemical composition and template capacity of chromatin for *in vitro* RNA synthesis has been described elsewhere (Spelsberg *et al.*, 1971b).

organs were first determined. The antigenic properties of the nonhistone protein–DNA complexes were then compared with those of the corresponding native (intact) chromatins. Antibodies were produced in rabbits against a preparation of oviduct nucleoacidic protein (acidic protein fractions AP_3 and AP_4 complexed to DNA) which was prepared from oviduct chromatin of chicks which had been injected with DES for 15 days. The method of quantitative complement fixation as described by Wasserman and Levine (1961) was chosen for testing the antigenic properties.

First the quantitative complement fixation curve for the antigen (i.e., chick oviduct nucleoacidic protein from 15 days stimulated chicks) used for immunization of the rabbits was determined. As can be seen in Fig.

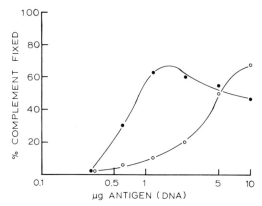

FIG. 5. Complement fixation by varying quantities of chick oviduct chromatin (○) and nonhistone protein–DNA complex (●) prepared as described in the text in the presence of antisera against nonhistone protein–DNA complex. Nucleoacidic protein from chick oviduct prepared as described in the text was used for immunization of rabbits. The preparation was stored at −20°C, thawed and homogenized in a Teflon-pestle glass homogenizer before use. Equal volume of complete Freund's adjuvant was added, and the mixture was homogenized again; 110 μg of nonhistone protein (185 μg of DNA) was injected into male New Zealand rabbit toe pads (hind feet) once weekly for 2 weeks, and for 6 more weeks intramuscularly. The rabbits were bled by cardiac puncture 7 days after the last injection, and the resulting serum (diluted 1:400) was used for determination of microcomplement fixation by varying amounts of different antigens by the method of Wasserman and Levine (1961). The anticomplementarity (binding of the complement in the absence of antiserum) was tested in the whole range of concentration used and subtracted from the total complement fixation.

5, complement fixation indicates the presence of antibody reacting with the above antigen. Figure 5 also shows comparison of the complement fixation curve for the intact chromatin isolated from the same tissue source. It can be seen that about 5-fold more chromatin was needed in comparison with acidic protein–DNA complex to produce the same percent of fixation. These data indicate that in native chromatin only 20% of the antigenic sites belonging to the acidic protein–DNA complexes are open, i.e., accessible for the antibody. It is probable that the rest of these antigenic sites are covered by histones and therefore are not reacting with the antibody.

The tissue specifity of the nucleoacidic protein fraction was then studied. Nucleoacidic protein was isolated from the chromatins of chick heart, liver, and spleen and tested for antigenicity using the same antiserum prepared against oviduct nucleoacidic protein described above. Results in Fig. 6 show that the nucleoacidic proteins from other organs show

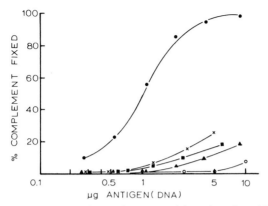

FIG. 6. Complement fixation by various quantities of nucleoacidic protein from chick oviduct and other organs in the presence of rabbit antiserum. ●, Oviduct from 15-day-treated chicks; ■, liver, ▲, spleen, X, heart, and ○, oviduct from untreated chicks.

a very limited affinity for this antibody. This limited affinity suggests that a large number of the antigenic sites present in oviduct preparations are tissue specific. These results indicate a considerable dissimilarity in antigenic sites of the acidic protein–DNA complexes which are present in the chromatins of different organs. Finally, Fig. 6 shows that preparations of nucleoacidic proteins from undifferentiated oviduct, i.e., oviducts of unstimulated chicks, show very little antigenicity under the conditions of the complement fixation assay.

Figure 7 shows changes in the antigenicity of the nucleoacidic proteins (containing AP_3 and AP_4) during growth and differentiation of chick oviduct. Fixation of complement by nucleoacidic proteins, which were isolated from oviducts of chicks injected with DES for various periods of time, show a gradual appearance of antigenicity with length of estrogen administration. When measured in the presence of antisera against nucleoacidic proteins of 15-day-stimulated animals the chromatin preparations from oviducts of 12-day-stimulated chicks fix complement to approximately the same extent as those from 15 or 19 days, respectively. This is in accordance with the results published earlier that the differentiation of chick oviduct by DES is completed after 12 to 15 days of estrogenic stimulation (O'Malley et al., 1969). Thus development of antigenicity of the nucleoacidic proteins show a developmental change which coincides with the morphological development of the organ.

In conclusion: (a) the nucleoacidic proteins isolated from chromatin complexed to DNA are good immunogens which give complement-fixing

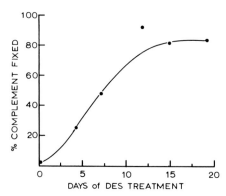

FIG. 7. Changes of antigenicity of nonhistone protein–DNA complexes during growth and differentiation of chick oviduct. Nucleoacidic protein complexes (10 μg of DNA per assay) were isolated as described in the text from oviducts of chicks which were administered diethylstilbestrol (DES, 5 mg daily) for various periods of time. Antiserum (dilution 1 : 400) against nonhistone protein–DNA complexes prepared from oviducts of chicks injected with DES for 15 days was used in all experiments.

antibody; (b) the antibodies do react strongly with the preparations from the homologous organ (chick oviduct), whereas the affinity for preparations of nucleoacidic proteins from other organs (liver, heart, spleen) is very low; this is indicative that the arrangement of the antigenic sites inherent to the acidic proteins in chromatin is organ specific; (c) when antibodies are produced against preparations not containing histones, they still react with the intact chromatin from the homologous organ, but with four to five times less affinity; this is an indication that about 20% of the antigenic sites belonging to the nucleoacidic protein are exposed and accessible to antibody in native chromatin containing histones; (d) during development of chick oviduct the antigenic sites for acidic proteins undergo marked alterations which involve either changes in protein species or structural alterations of already existing proteins.

B. Circular Dichroism

Structural analysis of the protein–DNA complexes were carried out using circular dichroism (CD) analysis under standardized conditions (Fig. 8). Data are expressed as changes in the magnitude of mean residue ellipticity at 275 nm which primarily reflect changes in DNA conformation. Figure 8 shows composite data concerning the effects of estrogen on several chemical and physical properties of oviduct chromatin and oviduct nucleoacidic protein fraction. The values for template capacity

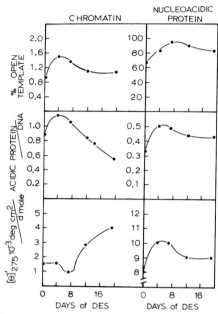

FIG. 8. Comparative properties (% open template, acidic protein:DNA ratio, and circular dichroism ellipticity at 275 mM) of chick oviduct chromatin and nucleoacidic protein as a function of days of DES stimulation. The percent open template was assayed using *Escherichia coli* RNA polymerase using methods described elsewhere (Spelsberg et al., 1971a). Acidic protein is that fraction of chromatin protein remaining after histone removal as described elsewhere (Spelsberg et al., 1971b). Mean residue ellipticity at 275 nm ([θ]' 275) is based on DNA concentration using a value of 309 as the mean residue weight. A Cary Model 60-CD using 1 mm cells at 2°C was used in each instance.

and acidic protein composition of the chromatin shown in the top and middle panels of both columns were taken from Table 1 and have been discussed above. The lower panels show results of the circular dichroism studies. The left column represents results of studies with intact chromatin whereas the right column presents data observed with the nucleoacidic protein fraction. The magnitude of the ellipticity at 275 nm for the chromatins is much less than that for pure DNA at all stages of development. DNA complexed with proteins as in chromatin has previously been shown to have a reduced positive band at 275 nm (Simpson and Sober, 1970; Wagner and Spelsberg, 1971; Fasman et al., 1970).

This decreased ellipticity has been suggested to be caused by an altered geometry of the DNA when complexed with proteins. However, as development of the oviduct progresses, the chromatin DNA displays a gradual increase (after 8 days) in the magnitude of ellipticity (Fig. 8). Other studies suggest that this increase in ellipticity may represent an "opening" of the DNA, i.e., removal of proteins from increasing areas of the DNA (Wagner and Spelsberg, 1971). The striking differences in the circular dichroic spectral pattern at 275 nm and the change in the percent open template as measured by the ability of the chromatin to serve as a template for RNA synthesis may be due to the inability of the bacterial RNA polymerase to transcribe selective DNA regions in chromatin. This inability may be due to steric interference or lack of proper initiation sites on the exposed DNA. In the studies with the nucleoacidic proteins, the magnitude of ellipticity at 275 nm follows a pattern throughout oviduct development which correlates with that of the percent open template. The nucleoacidic protein shows a much greater magnitude of ellipticity at 275 nm than the chromatin at all stages of oviduct development. This suggests that the chromatin bound proteins dictate DNA conformation. Although the precise origin of the conservative DNA spectra is unknown, recent calculations based on base–base interactions of polynucleotides indicate $\pi \rightarrow \pi^*$ transitions arising from a parallel base conformation in the B form of DNA as the principal contributor to the characteristic dichroic spectra (Johnson and Tinoco, 1969). The removal of chromatin-bound proteins appears to allow these transitions to occur progressively, as a result of conformational change. Moreover, the data presented suggest that the acidic proteins promote the B type structure during DES stimulation and that the histone moiety maintains this structure at the end of DES stimulation.

The low UV spectra of chromatin and nucleoacidic protein reflect primarily protein conformation with a secondary influence from DNA (Fig. 9). In chromatin it is reasonable to assume that the contribution from DNA is negligible since the 275-nm band is greatly reduced from that of native DNA. Thus the sizable shift in the 19-day, chromatin-negative, 210-nm peak as illustrated in Fig. 9 represents protein conformational changes. Based on the calculations of Greenfield and Fasman (1969), there is <10% α-helix prior to day 19 with a progressive increase to ~20% on day 19. Conformational changes also seem to be reflected in the nucleoacidic protein (Fig. 9). However, the DNA contribution here would be expected to be greater than that of chromatin, so that the origins of these changes is obscure at this time.

Thus the CD analysis of chromatin during estrogen administration supports the concept that a major alteration of the steric conformation of target cell chromatin occurs during this hormone-induced differentiation. Coupled with the data showing major quantitative and qualitative changes in the nucleoacidic proteins of chromatin, these studies provide

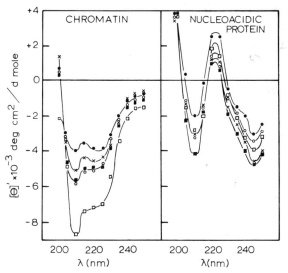

Fɪɢ. 9.   Low UV circular dichroic spectra of chromatin and nucleoacidic protein of chick oviduct as a function of days of diethylstilbestrol (DES) stimulation. The mean residue ellipticity was calculated on the mole fraction of protein and mole fraction of DNA present in each sample using a value of 115 for the average protein residue weight and 309 for the average DNA residue weight. All runs were in 1-mm cells at 2°C using an instrument full range of 0.04 or 0.1 degree and a constant VISA bandwidth.

strong evidence that differentiation represents progressive alterations in chromatin biochemistry which may result in changes in cell structure and function.

V. MESSENGER RNA TRANSLATION

The data presented in the preceding section are further confirmation of the premise that major changes occur in the transcriptional apparatus during development of the oviduct. It is also apparent that the differentiation process results in the synthesis of new organ-specific proteins, such as ovalbumin and avidin (O'Malley *et al.*, 1969; Means *et al.*, 1971). Thus one would expect that new species of messenger RNA are transcribed from the oviduct genome during development. However, the only direct proof of messenger RNA is the ability of an isolated RNA function to direct the synthesis of a specific protein in a chemically defined cell-free system. We have, therefore, initiated studies designed to develop a ribosomal system capable of translating exogenous message with fidelity in order to assess changes in messenger RNAs which occur during estrogen-mediated oviduct development.

We have previously described the isolation and characterization of a polyribosomal system from the oviduct of the estrogen-stimulated chick which is capable of cell-free protein synthesis (Means et al., 1971). This system is capable of synthesizing and releasing polypeptides in a time-dependent manner. A portion of the released peptides are immunologically similar to the tissue-specific protein ovalbumin (Means and O'Malley, 1971). It was recognized that this synthesis may have involved only the addition of a few amino acids to previously existing polypeptide chains; that is, ovalbumin chains may not have been initiated *in vitro*. Furthermore, the polysomal system did not respond to addition of the exogenous messenger polyuridylic acid. It has been demonstrated that washing bacterial and rabbit reticulocyte ribosomes with high salt yields a fraction containing protein factors which are required for the initiation of polypeptide synthesis *in vitro* (Lengyel and Söll, 1969; Prichard et al., 1970). Therefore, we began to further define our protein synthesis system in order to prepare materials capable of both initiating protein synthesis and translating natural endogenous message with fidelity and also capable of translating exogenous mRNA.

Polyribosomes prepared from the estrogen-stimulated chick oviduct were washed with a solution containing 0.25 M sucrose, 0.1 mM

TABLE 2

PROTEIN SYNTHESIS ON SALT-WASHED RIBOSOMES FROM OVIDUCT[a]

Incubation conditions	Natural message system (cpm [^{14}C]Phe/mg protein)	Polyuridylic acid system (cpm [^{14}C]Phe/mg protein)
Complete	400	3,100
Complete + KCl-wash	3400	29,300

[a]Polysomes were isolated from oviducts of chicks treated for 15 days with DES and resuspended in standard sucrose (0.25 M sucrose, 0.001 M dithiothreitol, 0.001 M EDTA) pH 7.0, to a concentration of 120 $A_{260\,nm}$/ml. KCl was then added to a final concentration of 0.5 M, and samples were centrifuged at 150,000 g for 1 hour. The salt wash was decanted and stored at $-70°C$ in 200-μl aliquots. Pellets were resuspended in standard sucrose to a final concentration of 120 $A_{260\,nm}$/ml. Ribosomes (110 μg of ribosomal RNA:1,2 $A_{260\,nm}$) were incubated in the complete-cell-free system for 20 minutes at 37°C. Concentrations of constituents in a final volume of 0.1 ml were as follows: ATP, 1.0 mM; GTP, 0.5 mM; PEP, 30 mM; pyruvate kinase, 1.5 IU; Mg^{2+}, 5.0 mM; K$^+$, 100 mM; dithiothreitol, 1.0 mM; Tris-HCl, 30 mM (pH 7.2 at 25°); 19 unlabeled amino acids, 20 μM each; [^{14}C]phenylalanine, 20 μM (0.1 μCi); and AS$_{70}$ fraction, 180 μg protein. Polyuridylic acid and KCl wash when added were present in amounts of 10 μg and 70 μg of protein, respectively. Following the incubation period reactions were terminated by addition of 1.0 ml of 10% trichloroacetic acid and prepared for determination of radioactivity as previously described (Means et al., 1971).

Na$_2$EDTA, pH 7.0, 1.0 mM dithiothreitol, and 0.5 M KCl as first reported by Miller and Schweet (1968) for the rabbit reticulocyte. The suspension was separated into ribosomes and a supernatant fraction by centrifugation at 105,000 g. Table 2 demonstrates that (a) the addition of the KCl wash stimulates protein synthesis in the natural message system; (b) the salt-washed ribosomes now respond to poly(U); and (c) the KCl wash stimulates the polymerization of phenylalanine in the poly(U)-dependent system. The protein nature of the KCl wash has been demonstrated by its sensitivity to heat, Pronase, and N-ethylmaleimide, but not to ribonuclease (Comstock et al., 1972).

The KCl supernatant fraction, which we have labeled AvF for its avian origin, was found to be active in stimulating both the rate and the extent of polymerization when compared to the system incubated with saturating amounts of the 30 to 70% ammonium sulfate fraction (AS-70) containing transfer factors and aminoacyl synthetases (Fig. 10). It can be seen that incorporation in the complete system is linear for 30 minutes.

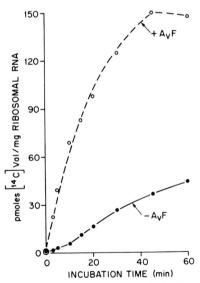

FIG. 10. Time course for the incorporation of [^{14}C]valine into protein. A 100-μl aliquot was removed at each time point from a 1.0 ml reaction mixture. The reaction mixture contained 1.10 mg of ribosomes (as RNA), 700 gm of AvF (as protein), and 1.8 mg of AS-70 enzyme fraction (as protein). The remaining components were added to final concentrations as described in Table 2. A zero time blank of 1.1 pmoles was subtracted from each point. One pmole of [^{14}C]valine is equivalent to 572 dpm.

TABLE 3

THE COMPLETE REACTION MIXTURE CONTAINED 110 μG OF RIBOSOMAL
RNA (1.4 A_{260} UNITS), 70 μG OF AvF PROTEIN, AND OTHER
COMPONENTS DESCRIBED IN TABLE 2[a]

Ribosomes (R)	Ribosomal wash fraction (AvF)	Supernatant enzyme fraction	pmoles [^{14}C]Val incorporated
R	(AvF)	AS_{70}	15.8
R	—	—	0.3
—	(AvF)	—	0.1
—	—	AS_{70}	0.1
—	(AvF)	AS_{70}	0.1
R	—	AS_{70}	2.4
R	(AvF)	—	9.5

[a] Supernatant enzyme fractions (containing transfer factors and aminoacyl synthetases) were prepared as described previously and added in saturating amounts as determined in separate experiments. The crude cell sap (150,000 g, R_{av}) routinely inhibited incorporation in all amounts tested. Incubation was for 20 minutes at 37°.

We have also determined that the cell-free ribosomal system exhibits linear dependence upon both the numbers of ribosomes and the amount of AvF protein. Finally this protein-synthesizing system exhibits the usual requirements for maximal polymerization activity, that is, the AvF-dependent reaction is highly dependent upon ATP, an energy-generating system and Mg^{2+}.

Dependencies of the system are shown in Table 3. The complete system containing 110 μg of ribosomal RNA, 70 μg of AvF protein, and 180 μg of AS-70 protein incorporated 15.8 pmoles of [^{14}C]valine. Less than 0.3 pmole as incorporated with ribosomes alone, and less than 0.1 pmole with the AvF fraction, the AS-70 fraction, or both of these together. The system with the ribosomes and the AS-70 fraction incorporated 2.4 pmoles, or 15% of the activity of the complete system. Ribosomes and AvF fraction alone incorporated 9.5 pmoles, or 60% of the activity of the complete system. This would indicate that AvF fraction is not saturating for transfer factors and/or aminoacyl synthetases.

The Mg^{2+} optimum for the AvF-dependent system was of critical importance since a lowering of the Mg^{2+} optimum has been associated with chain initiation and fidelity of translation in both bacterial and reticulocyte systems (Lengyel and Söll, 1969; Prichard et al., 1970). In our system utilizing endogenous natural message and [^{14}C]valine incorporation, the Mg^{2+} optimum in the absence of the AvF fraction is 7 mM (Fig.

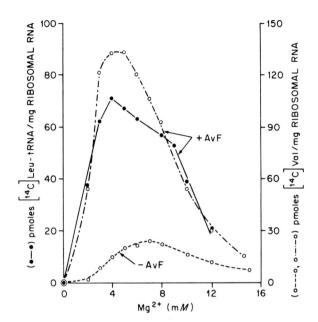

Fɪɢ. 11. [^{14}C]Valine and [^{14}C]Leu-tRNA incorporation into protein as a function of Mg^{2+} concentration. Each reaction mixture contained the appropriate components for the different assay conditions as described previously (Means *et al.*, 1971). Blanks of 1.3 pmoles of [^{14}C]valine and 1.8 pmoles of [^{14}C]leu-tRNA obtained for each assay condition in the absence of Mg^{2+} were subtracted from the appropriate points.

11). In the presence of the AvF fraction, there is a lowering of the Mg^{2+} requirement to 4 m*M*. It can be seen that AvF stimulates both rate and extent of protein synthesis using [^{14}C]leucyl-tRNA as the radioactive substrate. Moreover the lower Mg^{2+} optimum is also found when [^{14}C]leucyl-tRNA is used as substrate. This indicates that the acylation of tRNA is neither responsible for the AvF stimulation of protein nor for the lower Mg^{2+} optimum.

The next problem was to determine whether the AvF fraction was distinct from transfer factors. For these experiments we utilized natural message-deficient rabbit reticulocyte ribosomes and partially purified reticulocyte elongation factors, T1 and T2. This assay employs the poly(U)-directed polymerization of [^{14}C]Phe-tRNA assayed in the presence of saturating amounts of T1 and T2. Again, AvF fraction stimulates the rate of synthesis and lowers the Mg^{2+} optimum from 10 m*M* to 6 m*M* (Fig. 12). These data then, demonstrate that the stimulatory proteins present in AvF fraction are distinct from T1 and T2.

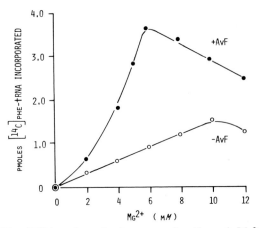

Fig. 12. [^{14}C]Phe-tRNA polymerization as a function of Mg^{2+} concentration. Each 50 μl reaction mixture contained 0.2 A_{260} of twice washed (0.5 M KCl) rabbit reticulocyte ribosomes (incorporation was linear to 0.35 A_{260} unit), 0.5 A_{260} unit of poly(U), 5.5 pmoles of chick oviduct [^{14}C]Phe-tRNA (840 cpm/pmole), and 30 μg of AvF protein. Saturating amounts of partially purified rabbit reticulocyte T1 protein (20 μg) and T2 protein (3.0 μg) were used in all tubes. The reaction mixture contained Tris · HCl (pH 7.2), 30 mM; GTP (pH 7.0), 0.5 mM; phosphoenol pyruvate (pH 7.0), 3.0 mM; pyruvate kinase, 1.5 EU.; dithiothreitol, 1.0 mM; and KCl, 100 mM (partially supplied by AvF fraction). Incubation was for 2 minutes at 37°C. A blank of 0.24 pmole obtained in the absence of Mg^{2+} was subtracted from each point. Incorporation without the addition of poly(U) was 0.10 pmole in the presence of AvF and 6 mM Mg^{2+}.

In order to demonstrate fidelity of translation in a cell-free system, one must show that this system is capable of *de novo* synthesis of a specific protein molecule. Figure 13 is an outline of the initial steps that we employ in an attempt to analyze the product synthesized in the AvF-dependent natural message system. A large reaction mixture is incubated and then separated into ribosomes and supernatant by ultracentrifugation. Incorporated radioactivity is determined in aliquots of each fraction. Under these conditions 27% of the incorporated counts are released into the supernatant. The supernatant is then passed through an affinity column to which highly specific antiovalbumin is coupled (Cuatrecasas, 1970). The column is washed thoroughly and then eluted with 6 M guanidine · HCl. Fractions are collected, pooled, and dialyzed. It can be seen that 14% of the released peptides are immunologically similar to ovalbumin.

Thus, we have described the properties of a new protein synthesis system derived from a hormone-sensitive target tissue. The system is depen-

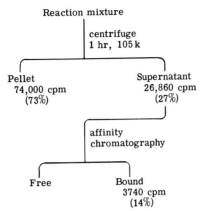

FIG. 13. AvF-dependent synthesis of immunologically reactive ovalbumin on oviduct ribosomes *in vitro*. The reaction mixture (200 μl) on which the analysis was carried out contained 2.0 mg of ribosomal RNA (20.0 A_{260} units), 400 μl of AvF (1.2 mg of protein), 3.6 mg of AS-70 protein, and other components as described in Table 1. After incubation at 60 minutes at 37°C, the reaction mixture was centrifuged at 105,000 g (R_{av}) for 60 minutes. The supernatant was carefully removed, duplicate aliquots were prepared for counting (Means *et al.*, 1971), and the remainder were applied to an antiovalbumin affinity column. The bound counts were collected in 2 ml after elution with 6 M guanidine · HCl and dialyzed against H_2O; 1 ml was counted in Spectroflor (Amersham/Searle):toluene:triton X-100 (42 ml:958 ml:500 ml). The ribosomal pellet was resuspended in H_2O and counted as hot acid-insoluble material.

dent upon a high salt wash (AvF) of oviduct polyribosomes. The system shows linear kinetics for time, ribosomes, and the AvF fraction. The AvF fraction lowers the Mg^{2+} optimum of the reaction under a variety of assay conditions. Finally, 14% of the released peptides synthesized in this system have been shown to possess ovalbumin immunoreactivity. Our data would suggest that this AvF-dependent oviduct system is capable of chain-initiation and a high degree of fidelity in translation. We hope to utilize this system to investigate directly changes in the content of specific mRNA's which occur during estrogen-mediated differentiation of the oviduct.

The chick oviduct then appears to be a suitable model system in which to study in depth the mechanism of estrogen-induced development and differentiation. Future studies will attempt to more closely coordinate the initial stages of cell proliferation with changes in specific restriction of nuclear chromatin and RNA transcription and eventually understand these nuclear events in relation to cytoplasmic translation of newly synthesized messenger RNA.

REFERENCES

CHYTIL, F., and SPELSBERG, T. C. (1971). Tissue differences in antigenic properties of nonhistone protein-DNA complexes. *Nature (London) New Biol.* **233**, 215–217.

COHEN, S., O'MALLEY, B. W., and STASTNY, M. (1970). *In vivo* and *in vitro* induction of ornithine decarboxylase by estrogens. *Science* **170**, 336–338.

COMSTOCK, J. P., O'MALLEY, B. W., and MEANS, A. R. (1972). Stimulation of cell free polypeptide synthesis by a protein fraction extracted from chick oviduct polyribosomes. *Biochemistry* **11**, 646–652.

CUATRECASAS, P. (1970). Protein purification by affinity chromatography. *J. Biol. Chem.* **245**, 3059–3065.

FASMAN, G. D., SCHAFFENHAUSEN, B., GOLDSMITH, L., and ADLER, A. (1970). Conformational changes associated with F-1 histone-deoxyribonucleic acid complexes. Circular dichroism studies. *Biochemistry* **9**, 2814–2822.

GREENFIELD, N., and FASMAN, G. D. (1969). Computed circular dichroism spectra for the evaluation of protein conformation. *Biochemistry* **8**, 4108–4126.

GURDON, J. B. (1962). Adult frogs derived from the nuclei of single somatic cells. *Develop. Biol.* **4**, 256–273.

HNILICA, L. S. (1967). Proteins of the cell nucleus. *Progr. Nucl. Acid Res. Mol. Biol.* **7**, 25–105.

JOHNSON, W. C., and TINOCO, I. (1969). Circular dichroism of polynucleotides: A simple theory. *Biopolymers* **7**, 727–749.

KOHLER, P. O., GRIMLEY, P. M., and O'MALLEY, B. W. (1969). Estrogen-induced cytodifferentiation of the ovalbumin-secreting glands of the chick oviduct. *J. Cell Biol.* **40**, 8–27.

LENGYEL, P., and SÖLL, D. (1969). Mechanism of protein biosynthesis. *Bacteriol Rev.* **33**, 264–301.

McGUIRE, W. L., and O'MALLEY, B. W. (1968). RNA polymerase activity of the chick oviduct during steroid-induced synthesis of a specific protein. *Biochim. Biophys. Acta* **157**, 187–194.

MEANS, A. R., and O'MALLEY, B. W. (1971). Assessment of sex steroid action in vitro. *In* "In Vitro Methods in Reproductive Cell Biology," (E. Diczfalusy, ed.), pp. 318–336. Bogtrykkeriert Forum, Copenhagen.

MEANS, A. R., ABRASS, I. B., and O'MALLEY, B. W. (1971). Protein biosynthesis on chick oviduct polyribosomes. I. Changes during estrogen-mediated tissue differentiation. *Biochemistry* **10**, 1561–1570.

MILLER, R., and SCHWEET, R. (1968). Isolation of a protein fraction from reticulocyte ribosomes required for *de novo* synthesis of hemoglobin. *Arch. Biochem. Biophys.* **125**, 632–646.

OKA, T., and SCHIMKE, R. T. (1969a). Interaction of estrogen and progesterone in chick oviduct development. I. Antagonistic effect of progesterone on estrogen-induced proliferation and differentiation of tubular gland cells. *J. Cell Biol.* **41**, 816–831.

OKA, T., and SCHIMKE, R. T. (1969b). Interaction of estrogen and progesterone in chick oviduct development. II. Effects of estrogen and progesterone on tubular gland cell function. *J. Cell Biol.* **43**, 123–127.

O'MALLEY, B. W. (1967). *In vitro* hormonal induction of a specific protein. *Biochemistry* **6**, 2546–2551.

O'MALLEY, B. W. (1969). Hormonal regulation of nucleic acid and protein synthesis. *Trans. N.Y. Acad. Sci.* **31**, 418–503.

O'MALLEY, B. W., and McGUIRE, W. L. (1968a). Studies on the mechanisms of estrogen-mediated tissue differentiation. Regulation of nuclear transcription and induction of new RNA species. *Proc. Nat. Acad. Sci., U.S.* **60**, 1527–1534.

O'MALLEY, B. W., and McGUIRE, W. L. (1968b). Studies on the mechanism of action of progesterone in regulating the synthesis of specific proteins. *J. Clin. Invest.* **47**, 654–664.

O'MALLEY, B. W., ARNOW, A., PEACOCK, A. C., and DINGMAN, C. W. (1968). Estrogen-dependent increase in transfer RNA during differentiation of the chick oviduct. *Science* **162**, 567–568.

O'MALLEY, B. W., McGUIRE, W. L., KOHLER, P. O., and KORENMAN, S. G. (1969). Studies on the mechanism of steroid hormone regulation of synthesis of specific proteins. *Recent Progr. Horm. Res.* **25**, 105–160.

PERROTTA, C. A. (1962). Initiation of cell proliferation in the vaginal and uterine epithelia of the mouse. *Amer. J. Anat.* **111**, 195–204.

PRICHARD, P. M., GILBERT, J. M., SCHAFRITZ, D. A., and ANDERSON, W. F. (1970). Factors for the initiation of haemoglobin synthesis by rabbit reticulocyte ribosomes. *Nature (London)* **226**, 511–514.

SIMPSON, R. T., and SOBER, H. A. (1970). Circular dichroism of calf liver nucleodhistone. *Biochemistry* **9**, 3103–3109.

SOCHER, S. H., and DAVIDSON, D. (1971). 5-aminouracil treatment. A method for estimating G_2. *J. Cell Biol.* **48**, 248–252.

SOCHER, S. H., and O'MALLEY, B. W. (1973). Estrogen mediated cell proliferation during chick oviduct development and its modulation by progesterone. *Devel. Biol.* **30**, 411–417.

SPELSBERG, T. C., STEGGLES, A. W., and O'MALLEY, B. W. (1971a). Progesterone binding components of chick oviduct. III. Chromatin acceptor sites. *J. Biol. Chem.* **246**, 4186–4189.

SPELSBERG, T. C.., HNILICA, L. S., and ANSEVIN, A. T. (1971b). Proteins of chromatin in template restriction. III. The molecules in specific restriction of the chromatin DNA. *Biochim. Biophys. Acta* **228**, 550–562.

WAGNER, T., and SPELSBERG, T. C. (1971). Aspects of chromosomal structure. I. Circular dichroism studies. *Biochemistry* **10**, 2599–2605.

WASSERMAN, E., and LEVINE, L. (1961). Quantitative micro-complement fixation and its use in the study of antigenic structure by specific antigen-antibody inhibition. *J. Immunol.* **87**, 290–295.

II. Formation and Organization of Plant Cell Walls and the Plasma Membrane

Origin and Growth of Cell Surface Components

D. James Morré and W. J. VanDerWoude

Department of Botany and Plant Pathology and Department of Biological Sciences, Purdue University, Lafayette, Indiana

I. INTRODUCTION

Cell membranes and associated surface coats are complex structures with unique functions (Benedetti and Emmelot, 1968; Dowben, 1969). Together, they serve as barriers between a cell's interior and the external milieu; are loci of differential solute permeability, transport (both uptake and secretion), and impulse transmission; and are intimately involved in cellular processes of communication and recognition (Curtis, 1967; Davis and Warren, 1967; Mason, 1968). The cell surface is a primary site of hormone and virus receptors as well as other types of informational molecules which mediate cell–cell interactions and cell–environment interactions and which may be responsible for cell, organ, and tissue

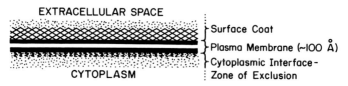

FIG. 1. Diagrammatic representation of the components of the cell surface.

specificity (Curtis, 1967; Davis and Warren, 1967; Mason, 1968; House and Wiedemann, 1970; Turkington, 1970). To effect physiological responses, all external stimuli must be transmitted in some way across the cell surface to responsive regions of the cytoplasm.

Cells have several distinct but interassociated regions at their surfaces: (1) the plasma membrane, (2) a surface coat external to the plasma membrane, and (3) an underlying differentiated layer of cytoplasm or zone of exclusion at the cytoplasm–plasma membrane interface (Fig. 1). Any or all of these surface components may be involved in the principal functions attributed to the cell surface, but attention is usually focused on the plasma membrane and its associated surface coats.

The elaboration and deposition of a surface coating rich in glycoproteins or polysaccharides (a glycocalyx = a sweet husk) is a common property of cells (Preston, 1959; Bennett, 1963; Rambourg et al., 1966; Revel and Ito, 1966). Higher plants and bacteria are familiar examples of sugar-coated cells, as are the chitinous exoskeletons of insects and crustaceans. Polysaccharide-rich coats are present at the external surface of animal cells. Even in cells where a surface coating is not obvious, evidence from electron microscopy and immunochemistry shows a polysaccharide-rich coating to be a general feature of cells. Once considered as extraneous, surface coats are involved in a myriad of cell interactions and adaptations, some of which may be particularly vital to the economy of differentiated cells organized into tissues.

Although the importance of the plasma membrane and surface coats has received considerable attention, the problem of the origin of these components of the cell surface is frequently overlooked. This paper summarizes new information from our laboratory on the origin of the cell membrane and surface coats as the product of integrated activities of the cell's internal membrane (endomembrane) system.

II. ORIGIN AND GROWTH OF CELL SURFACE COMPONENTS IN RAPIDLY ELONGATING POLLEN TUBES

One example of a developmental sequence in which elaboration of the cell surface involves concomitant addition of both plasma membrane and surface coat materials is provided by germinating pollen tubes of Easter lily (*Lilium longiflorum* L.). In nature, these pollen tubes transport the sperm cells down the long stylar column to the egg apparatus (Fig. 2a). After germination on the stigma, pollen tube growth through the style approaches a linear rate of 7.5 mm/hr in some species and is normally within the range of 1.5–3.0 mm/hr (Brewbaker and Kwack, 1964). This *in vivo* growth rate requires the impressive increase of nearly one pollen grain diameter (Fig. 2b) per minute. Pollen tubes offer a number of advantages for *in vitro* culture as well, requiring only oxygen, water, a suitable osmotic milieu, boric acid and a carbon source. Cultured *in vitro*, pollen tubes elongate at near linear rates of 0.1 to 0.3 mm/hr following germination. Although not as dramatic an increase as that of pollen tubes *in vivo*, this still represents an impressive growth rate for a single cell. Elongation of pollen tubes is almost exclusively a process of cell surface (membrane + surface coat) synthesis involving little or no increase in cytoplasm (Brewbaker and Kwack, 1964). This is attributed to continued vacuolation and plugging of mature regions of the tube (Brewbaker and Kwack, 1964; Alves *et al.*, 1968). As the tube elongates, plugs of callose (a β-1,3-linked glucan) are formed at regular intervals which separate the vacuolate, basal regions of cytoplasm from the organelle-rich, apical regions of cytoplasm. Surface growth is restricted to the pollen tube apex (Fig. 2c). In lily, phase microscope observations and the carbon particle method described by Rosen *et al.* (1964) show that tube elongation is restricted to a zone which extends back no more than 3–5 μm from the tube tip. This phenomenon of tip growth is characteristic of pollen tubes, fungal hyphae, rhizoids, and trichomes (including root hairs) (Grove *et al.*, 1970, and references cited therein).

A. Dual Organization of Plant Surface Coats

Surface coats of pollen tubes, like other types of plant cell walls, consist of two phases: a discontinuous microfibrillar phase and a continuous amorphous matrix in which the microfibrils are embedded. After refluxing isolated tube walls of germinated lily pollen (Figs. 2e,f) with ethanolamine to remove matrix polysaccharides, the microfibrillar phase is revealed as a "brush heap" of ca. 25 Å fibrils (Fig. 2g). The microfibrillar skeleton has been found throughout the tube wall including the very tip (Sassen, 1964). Data on microfibrillar dimensions, solubility, histochemical properties and glucose content are consistent with the fibers being α-cellulose, but their identification has not been rigorously defined.

The matrix phase appears amorphous in the electron microscope and consists of a complex mixture of very different types of polysaccharides

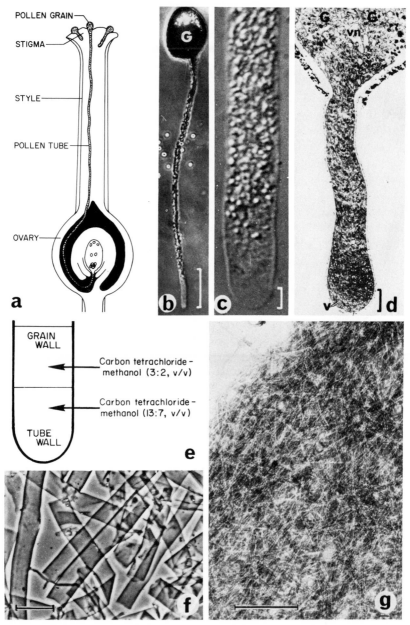

FIG. 2. Pollen tubes. (a) Diagrammatic representation of the growth of the tube of a germinated pollen grain from the point of germination on the stigma, through the style to the egg apparatus of the ovary. (b) Light micrograph of

plus proline- or hydroxyproline-rich (W. V. Dashek, personal communication), and perhaps other, cell wall proteins. The composition of matrix polysaccharides obtained from pollen tube walls by sequential extraction with hot water, hot 0.05 N H_2SO_4 and 4 N KOH (the standard pectin, acid-soluble hemicellulose and alkali-soluble hemicellulose fractions) is given by VanDerWoude et al. (1971). The residue, insoluble in 4 N KOH but soluble in 72% H_2SO_4 is the "cellulose" fraction. Matrix polysaccharides of the pollen tube, although rich in glucose (ca. 80%), contain significant amounts of arabinose, galactose, and galacturonic acid as well as mannose, xylose, rhamnose, and fucose. The separation methods used related to an artificial classification of cell wall polysaccharides (Albersheim, 1965), but the fractions obtained are sufficiently different to show a heterogeneous polysaccharide composition in the matrix phase (VanDerWoude et al., 1971). Galacturonic acid appears to be concentrated in polyuronides of the hot water-soluble polysaccharides but may be linked to other sugars in the hemicellulose fractions.

B. Surface Coat Formation by Vesicular Additions

The vesicular origin of the matrix polysaccharides and of the surface membranes of pollen tubes has been the subject of published reports (Rosen et al., 1964; Rosen and Gawlick, 1966; Sassen, 1964; Larson, 1965; Dashek and Rosen, 1966; Jensen and Fisher, 1970; VanDerWoude et al., 1971) (Table 1). Localized deposition of these materials involves the addition to the cell wall of membrane-bounded packets of materials in the form of secretory vesicles (Figs. 2d,3a,b). The vesicle membranes contribute new plasma membrane, and the vesicle contents provide precursors for the cell wall. These vesicles originate from the Golgi apparatus (Figs. 3b–d),

---

a pollen grain (G) germinated in culture. The length of the pollen tube illustrated (0.65 mm) is produced in culture in about 2 hours. Bar = 100 μm. (c) Nomarsky phase contrast micrograph of a pollen tube showing the apical growth region within the bar. Bar = 5 μm. (d) Electron micrograph of a germinated pollen grain (G) showing the vegetative nucleus (vn) and the apical accumulation of vesicles (V) of the emerging tube. From Sassen (1964). Reprinted by permission. Bar = 5 μm. (e) Nonaqueous procedure for separation of grain wall from tube wall of germinated pollen. (f) Phase contrast micrograph of an isolated pollen tube wall preparation prepared by the aqueous procedure of VanDerWoude et al. (1971). Bar = 50 μm. (g) Electron micrograph of a portion of a pollen tube wall fragment after extraction with ethanolamine (0.75 ml/mg) for 6 hours under reflux and negative staining with 2% potassium phosphotungstate. The hot ethanolamine extraction removed 80% of the cell wall material by weight and the microfibrillar skeleton (ca. 25 Å diameter microfibrils) of the ethanolamine-insoluble residue is shown. Bar = 0.2 μm.

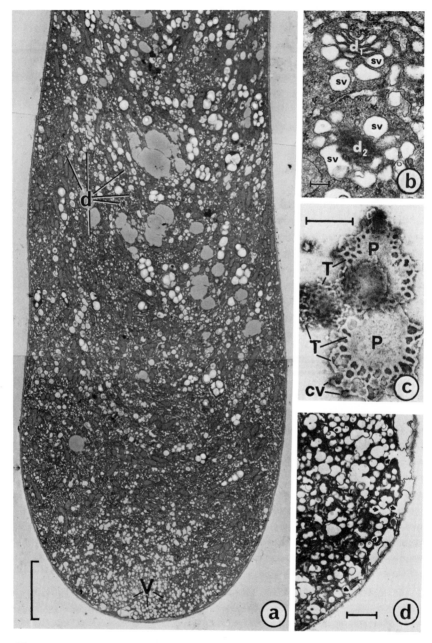

Fig. 3. Electron micrographs of the apical region of pollen tubes of *Lilium longiflorum*. Glutaraldehyde–acrolein–osmium fixation (VanDerWoude *et al.*, 1971). (a) Longitudinal section (approximately median) showing the distribution of cyto-

but the possibility that other cell components may be involved has not been excluded (VanDerWoude and Morré, 1968). The amount of vesicle membrane available for surface growth of the plasma membrane suggests a one-to-one stoichiometry between the amount of membrane contributed by vesicles and the increases in the plasma membrane during steady-state growth (Table 2).

The conclusion that secretory vesicles of the Golgi apparatus contribute significantly to the amorphous phase of the pollen tube cell wall is supported by a wide range of ultrastructural, histochemical, and biochemical studies (Table 1). Proof of this functional role has been aided by development of procedures for isolating secretory vesicles from pollen tubes by size discrimination of cytoplasmic components using Millipore filters (VanDerWoude et al., 1971). The isolated vesicles contain polysaccharides of composition similar to those of the hot water-soluble fraction of the cell wall (Table 3) (VanDerWoude et al., 1971).

In studies where radioactive glucose was supplied to the growth medium, the secretory vesicles reach constant specific activity in less than 1 hour while radioactivity accumulates in the cell wall. The vesicle fraction shows a specific activity and distribution of radioactivity after 1 hour which is similar to that of the hot water-soluble fraction of the cell wall after 10 hours of labeling (at which time sugars of the tube wall reach constant specific activity) (Table 4). These results establish a precursor–product relationship between the secretory vesicles and the amorphous or matrix phase of the cell wall (see also Mascarenhas, 1970) and show the direction of vesicle migration to be from the cytoplasm to the cell surface rather than in the reverse direction.

Thus, vesicular additions provide an attractive mechanism for the sur-

---

plasmic components. Secretory vesicles (V) are concentrated at the tube apex, the region of cell surface formation. The vesicles originate from dictyosome (d) located distal to the apex (montage of 3 electron micrographs). Bar = 5 μm. (b) Two adjacent dictyosomes (d) from the pollen tube cytoplasm 10–14 nm from the apex. A tangentially sectioned dictyosome (d_2) shows the numerous secretory vesicles (sv) attached to the central platelike portion of the cisternae via the system of peripheral tubules. Bar = 0.2 μm. (c) A dictyosome, isolated and negatively stained (Morré, 1970) from germinating lily pollen. The dictyosome is partially unstacked to show the central platelike region (P) and the system of peripheral tubules (T) as well as variations in cisternal features. The small cisterna near the forming face (top) is almost tubular, while the cisternae near the maturing face (left) are more platelike. Coated vesicles (CV) are attached to the cisternal tubules, which are a consistent feature of all dictyosome cisternae (Morré et al., 1971d). Secretory vesicles are not shown. Dictyosome-secretory vesicle connections are fragile, and most secretory vesicles are lost during isolation of dictyosomes from pollen tubes. Bar = 0.5 μm. (d) A portion of a pollen tube apex. Images of vesicle fusion (small arrows) are characteristic of this region. Bar = 1 μm.

TABLE 1

SUMMARY OF SOME EXPERIMENTAL OBSERVATIONS WHICH SHOW SECRETORY
VESICLES TO BE A SOURCE OF POLYSACCHARIDES OF THE CELL WALL
MATRIX DURING POLLEN TUBE GROWTH

Method of investigation	Results	Source[a]
Developmental electron microscopy	Ultrastructural patterns are those associated with vesicular additions to the cell wall	1–6
Isolation of vesicles and chemical characterization of vesicle contents	Similar to matrix polysaccharides of the cell wall	6, 8
[14C]Glucose incorporation and subsequent comparison of specific activities of constituent sugars among cell fractions	Sugars from isolated secretory vesicles reach constant specific activity in <1 hour while polysaccharides of cell walls reach constant specific activity much later	8
[3H]Inositol incorporation and electron microscope autoradiography	Silver grains deposited over tip cytoplasm and tube wall at growing tip	1
Electron microscope histochemistry based on differential staining	Similar staining of vesicle contents and tube wall	4–6, 8
Electron microscope histochemistry based on chemical and enzymatic extraction	Consistent with vesicle contents being cell wall matrix polysaccharides	1, 3
Light microscope histochemistry	Tip cytoplasm and tube wall show similar staining properties with some methods of cytochemical localization	1, 6
[3H]Inositol incorporation and light microscope autoradiography	Consistent with vesicle contents being precursors to cell wall matrix polysaccharides	7
Phase contrast and Nomarsky phase contrast light microscopy	Membrane flux at growing tip as required by surface growth via vesicular additions	8
Inhibitors	High CaCl₂ levels simultaneously reduce numbers of secretory vesicles and incorporation of sugars into matrix polysaccharides	8

[a] 1. Dashek and Rosen (1966). 2. Larson (1965). 3. Rosen and Gawlick (1966). 4. Rosen et al. (1964). 5. Sassen (1964). 6. VanDerWoude et al. (1971). 7. Young et al. (1966). 8. This report.

face growth of plant cells, the vesicle membrane contributing new plasma membrane and the vesicle contents providing cell wall materials. This type of cell surface deposition predominates in tip growing systems (Grove et al., 1970; Grove and Bracker, 1970) but fails to explain adequately the origin of surface coats in animal cells or secondary wall for-

TABLE 2

CALCULATED RATE OF VESICLE PRODUCTION AND
CELL SURFACE (MEMBRANE AND CELL WALL)
INCREASE FOR CULTURED POLLEN TUBES

Rate of tube elongation[a]	6 μm/min
Tube diameter[a]	16 μm
Wall matrix thickness[a]	0.05 μm
Vesicle diameter[a]	0.30 μm
Vesicle volume	0.014 μm^3
Vesicle surface	0.28 μm^2
Increase in wall volume	15 μm^3/min
Vesicle production	1075/min
Vesicle membrane production	300 μm^2/min
Increase in plasma membrane	300 μm^2/min

[a] Direct measurements. Remaining entries are calculated from these measurements.

TABLE 3

SUGAR COMPOSITION OF CELL WALL AND SECRETORY VESICLES

	Pollen tube cell wall[a]		
Sugar	Hot water soluble[b]	Hot water insoluble[b]	Secretory vesicles[c]
Galacturonic acid[d]	15.0	6.1	18.1
Glucose[d,e]	41.0	78.3	33.2
Galactose[d,e]	13.5	3.2	22.2
Mannose[d,e]	1.7	0.1	11.1
Arabinose[d]	19.7	9.1	5.1
Xylose[d]	5.0	1.1	6.1
Deoxyhexose[d,f]	4.1	2.1	4.2

[a] Isolated by the nonaqueous procedure (Fig. 2a).

[b] Extracted for 12 hours at 100°. This procedure extracts approximately 10% of the total dry weight of cell wall leaving 90% hot water insoluble.

[c] Isolated by Millipore filtration and differential centrifugation (VanDerWoude et al., 1971).

[d] Determined following separation by paper chromatography (procedure described by VanDerWoude et al., 1971).

[e] Determined as total hexose. Relative quantities of glucose, galactose, and mannose were calculated from data of VanDerWoude et al. (1971).

[f] Rhamnose plus fucose.

mation and late stages of primary wall formation in most higher plant cells (Northcote, 1971; Harris and Northcote, 1971). Here, vesicular additions of surface coats are observed infrequently or not at all.

TABLE 4

SPECIFIC ACTIVITY OF COMPONENT SUGARS COMPARING VESICLES
(1 HOUR LABELING PERIOD) AND THE HOT WATER-SOLUBLE
CELL WALL FRACTION (10 HOUR LABELING PERIOD)[a]

Sugar	Vesicle fraction (after 1 hour)		Hot water-soluble fraction (after 10 hours)	
	%	Cpm/mg	%	Cpm/mg
Hexose	68	1860	57	2000
Galacturonic acid	17	5090	15	2090
Pentose	11	4690	24	2400
Deoxyhexose	4	4620	4	3570

[a] Pollen was germinated 2 hours before labeling.

C. Extravesicular Formation of Surface Coat Materials

The compositional studies of isolated secretory vesicles from pollen tubes suggest that these cell components contribute much of the galacturonic acid to the cell wall, with significant contributions of pentoses, galactose, glucose, and deoxyhexoses. However, it seems unlikely that they are the sole contributors to the cell wall even in pollen tubes (i.e., insufficient glucose). Evidence for cell wall formation independent of secretory vesicles comes from studies in which pollen tube growth is inhibited by high concentrations of calcium chloride. These studies also provide confirmatory evidence for the role of secretory vesicles in the deposition of the polysaccharides of the wall matrix.

Elongation of pollen tubes is both stimulated and inhibited by calcium ions (Fig. 4a) with maximum inhibition at about 0.01 M calcium chloride. Other ions inhibit pollen tube elongation at high concentrations, but only the response to calcium was studied in detail.

Although pollen tube elongation was greatly suppressed by calcium ions, wall growth continued (Table 5). Tube length was inhibited by 70%, but wall synthesis was inhibited by only 45%. Thus, walls of the calcium-treated tubes were thicker than those of control tubes as verified by electron microscope observations (Fig. 5g,i). Calcium-inhibited pollen tubes were frequently characterized by apical swellings (Figs. 5d–f) and/or a spiraling growth pattern, the latter resulting in a zigzag appearance of the tubes (Figs. 5a–c).

Ultrastructurally, the pollen tubes were modified so that the zone of secretory vesicles at the tube tip was reduced (Figs. 5g,h; Table 6). The number of vesicles in the growing tip (0–4 μm from the tube apex) was

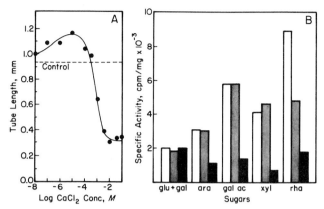

Fig. 4. (a) Effect of various concentrations of calcium chloride on growth of pollen tubes. Calcium chloride was added after 2.5 hours of germination. Tube length was measured after 4.5 hours of growth. (b) Specific activity of cell wall constituents of pollen tubes grown in the presence of [^{14}C]glucose for 17.5 hours ±0.01M CaCl$_2$. Included is a physiological control incubated for 10 hours in the absence of CaCl$_2$ (▦), in which an amount of wall growth occurred approximately equal to that of tubes grown in the presence of 0.01M CaCl$_2$ (■) (Table 5). □, 17.5-hour control grown without calcium chloride.

reduced by about 50%. In the region 4–8 μm from the tube apex, the number of vesicles was reduced even more dramatically by the calcium treatment while control values remained high (Table 6). The reduction in secretory vesicles was also reflected in the sugar content of secretory vesicle fractions isolated from calcium-inhibited and normal pollen tubes (Table 6).

When incorporation of radioactivity from [^{14}C]glucose in the presence and absence of 0.01 M CaCl$_2$ was examined, an informative pattern of inhibition was revealed (Fig. 4b). The first two bars are controls: a temporal control incubated for 17.5 hours and a physiological control (in which an amount of wall was produced comparable to that of the cal-

TABLE 5
WEIGHT OF TOTAL CELL WALL/LENGTH OF POLLEN TUBES

Growth criteria	Control (10 hours)	Control (17.5 hours)	0.01 M CaCl$_2$ (17.5 hours)
A. Average tube length (mm)	3.5	5.0	1.5
B. Mg cell wall/gm pollen	48	75	41
B/A	14	15	27

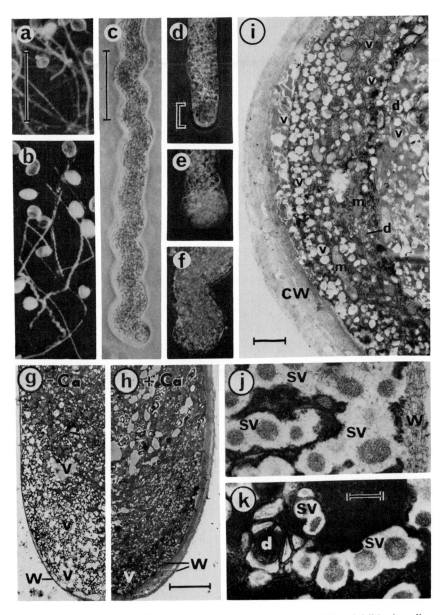

Fig. 5. (a–i) Structural comparisons of normal and calcium-inhibited pollen tubes treated as in Fig. 4a ±0.01 M CaCl$_2$. (a) Light micrograph (normal). (b) Light micrograph (0.01 M CaCl$_2$). Bar = 0.5 mm. (c,e,f) Phase micrographs (0.01 M CaCl$_2$). (d) Phase micrograph (normal). (c) Bar = 50 μm. (d,e,f) Bar = 10 μm. Pollen tubes grown in the presence of 0.01 M CaCl$_2$ are characterized by

TABLE 6

EFFECT OF CaCl₂ TREATMENT ON NUMBERS OF SECRETORY VESICLES AND
ON THE SUGAR CONTENT OF SECRETORY VESICLE FRACTIONS ISOLATED
FROM POLLEN TUBES OF *Lilium longiflorum*

Vesicles determined as:	$-CaCl_2$	$+CaCl_2$	% Ca inhibition
Vesicle profiles/10 μm^2 cytoplasm[a]			
0–4 μm from tube apex	57 ± 30	25 ± 8	56
4–8 μm from tube apex	44 ± 13	11 ± 8	75
μg Saccharide/gm pollen[b]	18	9	50

[a] After 5 hours of germination; the final 2 hours $\pm$ 0.01 M CaCl₂.

[b] Saccharide content of secretory vesicle fraction isolated after 3.5 hours of germination; the final 1 hour $\pm$ 0.01 M CaCl₂.

cium-inhibited tubes) incubated for 10 hours. With the exception of rhamnose, the wall constituents reach constant specific activity by 10 hours of labeling by [^{14}C]glucose. Most striking are the results for the hexose fraction (composed largely of glucose) which show no alteration of specific activity as a result of calcium inhibition. This is largely due to the fact that labeling of the cellulose fraction was unaffected or even stimulated by the calcium treatment. However, the constituents of the matrix polysaccharides (arabinose, galacturonic acid, xylose, and rhamnose) were inhibited to about the same extent as was wall growth and numbers of secretory vesicles. Thus, the studies with calcium inhibition not only show the expected relationship between vesicles and deposition of matrix polysaccharides, but provide clear evidence for a mechanism

a spiraling growth pattern or zigzag appearance (b,c,f), a bulbous swelling at the tube tip (e,f), or both (f). (g,h) Electron micrographs of near median sections of pollen tubes grown in the presence or absence of 0.01 M CaCl₂. Bar = 5 μm. The control tube in (g) shows numerous apical secretory vesicles (V) and the characteristically thin cell wall (W). The calcium-inhibited tube in (h) shows a marked reduction in the numbers of secretory vesicles and an abnormally thickened wall. (i) Electron micrograph of a near median section of a pollen tube treated with 0.01 M CaCl₂. The cytoplasm appears stratified and the zone of apical vesicles (v) is reduced to the point that mitochondria (m) and dictyosomes (d) are observed in the apical 2 μm of cytoplasm. Note thickened cell wall (CW). Bar = 1 μm. (j,k) Ultrastructural appearance of an unusual pollen tube in which normal migration of secretory vesicles has ceased and microfibrillar components resembling the microfibrils of the cell wall (W) have appeared within secretory vesicles (SV) of the apical cytoplasm in (j) and those associated with dictyosomes (d) in (k). These observations suggest that secretory vesicles contain a full complement of wall biosynthetic enzymes, but under normal circumstances only matrix polysaccharides are produced. Bar = 0.2 μm.

of glucan deposition independent of secretory vesicles. Recent studies by Dr. C. A. Lembi and W. J. VanDerWoude show that purified plasma membrane fractions from onion stem are active in the polymerization of glucose from UDP-glucose into a range of β-1,4- and β-1,3-linked glucans. These findings provide confirmatory evidence that glucan synthetases are present at the plant-cell surface.

D. Evidence for Sequential Biosynthesis of Surface Coats

Studies from cell-free systems show that polysaccharides are synthesized one sugar unit at a time by transfer from appropriate sugar nucleotide donors to acceptors which lack their full complement of sugar moieties (Hassid, 1969; Roseman, 1970). Glycosyltransferases catalyze each sugar addition, and specificity is determined by both the sugar nucleotide and the acceptor molecule. The glycosyltransferases which participate in elaboration of surface coats are particulate enzymes, i.e., bound to membranes and structured as part of the endomembrane system.

VanDerWoude et al. (1971) present cytochemical evidence for a spectrum of secretory vesicle types in pollen tubes ranging from just-forming secretory vesicles of the subapical cytoplasm with no detectable polysaccharides to mature secretory vesicles (polysaccharide content resembling that of the cell wall) at the tube apex. The vesicle membranes are elaborated at the Golgi apparatus, and the vesicles, in turn, appear to serve as organelles of polysaccharide synthesis or accumulation (VanDerWoude et al., 1969). Much of the polysaccharide content of the vesicles appears to be formed as the vesicles migrate. Progressive changes in the appearance of secretory vesicles associated with dictyosomes is commonly observed in plant cells (Mollenhauer and Whaley, 1963). Pickett-Heaps (1968) described an increase in the periodic acid–Schiff reaction of the contents of the vesicle lumens which progressed from the forming to the secreting face of dictyosomes. Manton (1966, 1967) and Brown et al. (1970) have demonstrated the progressive development of scales within dictyosome vesicles in species of scale-forming algae. Bonnett and Newcomb (1966) also suggested that the degree of polymerization of polysaccharides changes during vesicle migration. In general, surface coat biosynthetic activities correlate with the transformation of Golgi apparatus membranes from endoplasmic reticulumlike to plasma membranelike in the production of secretory vesicles (Morré et al., 1971d).

E. Composite Scheme for Cell Wall Formation in Pollen Tubes

From the work with pollen tubes and other plant cells (Mollenhauer and Morré, 1966; Northcote, 1968, 1971; Harris and Northcote, 1971;

VanDerWoude *et al.*, 1971), elaboration of surface coats seems to follow a temporal and spatial sequence in which polyuronides and a portion of the noncellulosic polysaccharides are synthesized in secretory vesicles of the Golgi apparatus. The bulk of the "cellulose" deposition occurs as a late event at the cell surface proper. A hypothetical sequence of polysaccharide biosynthesis is diagrammed in Fig. 6.

The suggestion that polyuronides are assembled within the secretory vesicles is based on the high galacturonic acid content of the vesicles (Table 4) and numerous cytochemical observations consistent with this view (Mollenhauer and Morré, 1966). A similar conclusion was reached by Dashek and Rosen (1966), who presented cytochemical evidence for the methyl esterification of free carboxyl groups in secretory vesicles during migration of the vesicles to the tube apex. Other noncellulosic polysaccharides of the wall matrix appear to be added in the later stages of vesicle migration, but biosynthesis of matrix polysaccharides by secretory vesicles is visualized as continuing even after the membranes have fused with the plasma membrane. "Cellulose" deposition *in vivo* appears to be largely confined to the cell surface (plasma membrane?).

Multiglycosyltransferase complexes have been postulated for both plants (Villimez *et al.*, 1968; Fig. 6) and animals (Roseman, 1970). Such

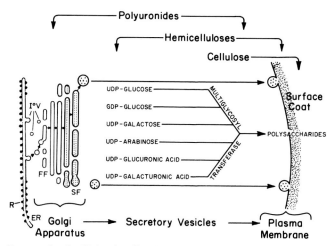

Fig. 6. Proposed subcellular localization of wall biosynthetic activities in pollen tubes superimposed upon a diagrammatic representation of Golgi apparatus functioning in membrane flow and differentiation adapted from Grove *et al.* (1970). The membrane-bound glycosyltransferase complex is according to Villimez *et al.* (1968). *ER*, Endoplasmic reticulum; *R*, ribosome; *FF*, forming face; *SF*, secreting face; *I°V*, primary vesicle. Arrows denote direction of membrane flow and vesicle migration.

complexes may be present on the membranes at the time of their formation at the rough-surfaced endoplasmic reticulum but function in various
phases of surface coat elaboration only when conformational controls or
substrate levels and availability dictate. Evidence that Golgi apparatus
and secretory vesicle membranes do contain the enzymes for synthesis
of the microfibrillar phase of the pollen tube wall is provided by the
unusual pollen tube of Figs. 5j,k (discussed in detail by VanDerWoude
et al., 1971).

F. Macromolecules Regulating Surface Coat Formation

One example of a specific control over the activity of a single glycosyltransferase complex is provided by studies with lactose synthetase (Table
7). This activity has been shown to be due to a single enzyme which
catalyzes the synthesis of N-acetylaminolactose in the presence of UDP-
galactose and N-acetylglucosamine as an acceptor but is inactive in the
synthesis of lactose with glucose as the acceptor (Brodbeck and Ebner,
1966; Brew, 1969). However, if a second protein, α-lactalbumin or specifier protein is added to the reaction, lactose is formed with glucose as
the acceptor. At the same time, the N-acetylaminolactose synthetic
capacity is reduced. Another example of control is provided by the studies
of Ordin and Hall (1968), where a glucan synthetase isolated from oat
coleoptiles synthesizes predominantly β-1,4-linked glucans at low UDP-
glucose concentrations but synthesizes predominantly β-1,3-linked
glucans at high UDP-glucose concentrations.

The relationship between the size of the apical zone of vesicles and
growth rate, as observed by light microscopy both for pollen tubes and
other tip growing systems (Grove *et al.*, 1970), suggest that vesicle pro-

TABLE 7

CONTROL OF DISACCHARIDE SYNTHESIS IN A GOLGI APPARATUS
FRACTION ISOLATED FROM RAT MAMMARY GLAND[a]

UDP-Galactose* + N-acetylglucosamine → N-acetylaminolactose*
UDP-Galactose* + glucose ⟶ lactose*
Specifier protein (= α-lactalbumin)

Exogenous specifier protein	Product synthesized (mµmoles/hr/mg/protein)	
	Lactose with glucose as acceptor	N-Acetylaminolactose with N-acetylglucosamine as acceptor
Absent	35	1457
Present	654	517

[a] From Keenan *et al.* (1970).

duction, vesicle coalescence at the growing tip and growth rate are closely coupled. Growth rate is necessarily tied to the rate of surface increase so a rapid incorporation of vesicles accompanies rapid growth, and vesicles do not accumulate extensively. With declining growth rate, the increase in extent of the apical zone (Alves *et al.*, 1968; Grove *et al.*, 1970) may reflect a decreased frequency of vesicle fusion. Yet, it is not known whether frequency of vesicle incorporation and/or production control growth rate or whether growth rate influences the rates of vesicle production and/or incorporation. In spite of this latter uncertainty, environmental factors which alter growth rates are expressed ultimately in an altered rate of vesicle production. Equilibrium between activities of cytoplasmic components and activities of components at the cell surface is achieved rapidly in tip-growing cells, indicating some form of feedback regulation which modulates vesicle production so as to keep pace with the requirements for surface growth.

III. ORIGIN AND GROWTH OF THE PLASMA MEMBRANE

Evidence from pollen tubes and other tip-growing plant cells suggests that in these cells the entire plasma membrane is derived via the Golgi apparatus–secretory vesicle export route. Thus, not only do secretory vesicles perform an important function in surface coat formation but they provide an important source of new plasma membrane to growing cells. The phenomenon whereby Golgi apparatus produce secretory vesicles whose membranes fuse with and contribute to the plasma membrane is not restricted to tip-growing cells, but occurs in a variety of cell types, both growing and nongrowing (Morré *et al.*, 1971d).

The ultimate source of Golgi apparatus membrane appears to be the rough-surfaced endoplasmic reticulum or nuclear envelope (Morré *et al.*, 1971d, and references cited therein). Individual stacks of Golgi apparatus cisternae (the dictyosomes) have a clearly defined forming face where new cisternae arise by fusion of small endoplasmic reticulum- or nuclear envelope-derived, primary vesicles. Membranes at this face resemble those of endoplasmic reticulum on the basis of thickness, staining characteristics, and cytochemical properties as determined by electron microscopy. At the opposite or secreting face, cisternae are lost through the production of the secretory vesicles (Mollenhauer, 1971). The membranes of the "mature" Golgi apparatus cisternae and those of the secretory vesicles are plasma membranelike and capable of fusing with the plasma membrane. These characteristics are essential to the functional role of secretory vesicles as vehicles for the transport of materials destined for export to the cell surface. Thus, membranes of secretory vesicles of the Golgi apparatus appear to be derived from endoplasmic reticulum or nuclear membrane, transformed during passage through the Golgi apparatus and destined to become plasma membrane. According to this concept, membranes are transferred and transformed along a chain of cell components in a subcellular developmental pathway. These dynamic concepts of Golgi apparatus functioning are illustrated in Fig. 6.

A. The Endoplasmic Reticulum–Golgi Apparatus–Secretory Vesicle–Plasma Membrane Export Route: A Discontinuous System of Plasma Membrane-Generating Transition Elements

If Golgi apparatus membranes are derived from endoplasmic reticulum membranes and are destined to become plasma membrane, they must be converted from endoplasmic reticulumlike to plasma membranelike somewhere en route. As summarized in Fig. 6, morphological evidence suggests that this conversion takes place at the Golgi apparatus (Grove et al., 1968; Morré et al., 1971b,d). Differential staining of Golgi apparatus membranes (including membranes of rat liver) show a gradual progression of membrane thickness and stainability from endoplasmic reticulumlike to plasma membranelike across the stacks of cisternae. Additionally, the gradient in membrane morphology from endoplasmic reticulumlike to plasma membranelike across the Golgi apparatus reflects a gradient in chemical and enzymatic composition among these cell components. Biochemical markers for mammalian plasma membranes and endoplasmic reticulum are encountered in Golgi apparatus fractions of both rat liver and mammary gland at concentrations intermediate between those of endoplasmic reticulum and plasma membrane. These findings have been summarized (Morré et al., 1971a,b,d; Keenan et al., 1973) and will not be repeated here.

Particularly critical to the concept of membrane flow are the numbers and identities of membrane proteins which characterize each of the cell components involved. Our studies (Yunghans et al., 1970; D. Bailey et al., results unpublished) using polyacrylamide- and starch-gel electrophoretic separations show rough- and smooth-surfaced endoplasmic reticulum membranes to be qualitatively and quantitatively identical, and similar to Golgi apparatus membranes. Each fraction (rough- and smooth-surfaced endoplasmic reticulum and Golgi apparatus) have the same numbers of major protein bands in the same positions on the gels. Although plasma membrane profiles have several bands in common with Golgi apparatus and endoplasmic reticulum, the membrane proteins appear to be more extensively glycosylated with attendant shifts in electrophoretic mobility. The extent to which differences in electrophoretic profiles between Golgi apparatus and plasma membrane (see also Fleischer and Fleischer, 1970) can be explained on the basis of progressive glycosylation in rat liver is currently under investigation. A striking correspondence between major protein bands comparing endoplasmic reticulum, Golgi apparatus, and plasma membrane from bovine mammary gland has been obtained in the laboratory of Professor T. W. Keenan at Purdue University.

The kinetics of membrane flow from endoplasmic reticulum to the plasma membrane via Golgi apparatus has been examined in short-time labeling and turnover studies with rat livers (Franke *et al.*, 1973). Purified cell fractions were extracted successively with 1.5 M KCl and 0.1% deoxycholate in 0.25 M sucrose and 0.01 M Tris, pH 7.6, with sonication to yield membrane proteins free from intravesicular, intracisternal, absorbed, and ribosome-associated proteins. As shown in Fig. 7, the order of labeling was (1) endoplasmic reticulum, (2) Golgi apparatus, and (3) plasma membrane. Rapid decreases in specific radioactivity followed maximal labeling of endoplasmic reticulum and Golgi apparatus membranes. These rapid turnover components are sufficient to account for

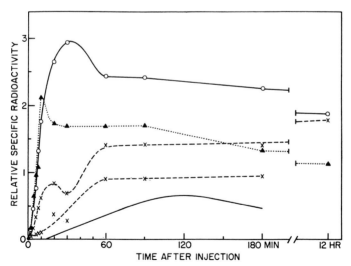

FIG. 7. Kinetics of incorporation of radioactivity from L-[^{14}C]-guanidoarginine into membrane proteins. The lower curve for plasma membrane is calculated on the basis of a contribution of 20% of the specific activity of the rough-surfaced endoplasmic reticulum to attempt to correct for the initial incorporation phase. O——O, Golgi apparatus; ▲•••▲, microsomes (endoplasmic reticulum); ×---×, plasma membrane; ——, blood serum albumin. From Franke *et al.* (1971).

membrane flow. For example, in hepatocytes, the quantity of endoplasmic reticulum exceeds that of the Golgi apparatus to the extent that the rapid turnover component of the endoplasmic reticulum would account for the total incorporation of radioactivity into the plasma membrane and more than account for incorporation of radioactivity into the Golgi apparatus.

Morphological observations, taken together with the findings that Golgi apparatus membranes are morphologically and compositionally inter-mediate between endoplasmic reticulum and plasma membrane as well as the kinetics of labeling and turnover of membrane proteins are consis-

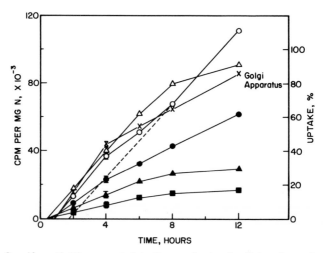

FIG. 8. Specific activities on a total nitrogen basis of cell fractions of onion stem from tissue incubated with 2μCi (168 μCi/mole) [U-¹⁴C]leucine. The experiment was carried out as described previously for other metabolites (Morré, 1970). The deter-minations at 4 hours were repeated on 3 different days ± standard deviation. Onions were used without prestorage. [¹⁴C]Leucine was taken up rapidly from the medium (△——△) following an apparent lag of 30–40 minutes. By 12 hours, more than 90% of the label had been removed from the external medium. After the lag in uptake, leucine incorporation into microsomes (endoplasmic reticulum) (▲——▲) and the nuclear fraction (■——■) was linear but plateaued as the medium was depleted of the isotope. Incorporation of leucine into dictyosomes and the plasma membrane fraction (○——○) lagged approximately 30 minutes to 1 hour behind incorporation into microsomes, whereas incorporation into mitochondria-proplastids (●——●) was intermediate between microsomes and dictyosomes both in specific activity and lag time (the apparent low specific activity of microsomes and nuclei results in part from their high nucleic acid contents since results are expressed on a nitrogen basis). Incorporation of radioactivity into plasma membrane was characterized by an initial phase coinciding with or preceding incorporation into dictyosomes and a second phase of incorporation between 4 and 12 hours which lagged (by extrapola-tion) approximately 30 minutes behind the initial linear phase of incorporation. ×——×, Golgi apparatus.

tent with the hypothesis that *membrane differentiation* coupled with *membrane flow* represent an important mechanism for the origin of plasma membrane from endoplasmic reticulum. However, as discussed in the section which follows, the Golgi apparatus is not to be considered as the only source of plasma membrane available to cells. Both morphological and kinetic evidence suggest alternative pathways.

B. Endoplasmic Reticulum–Plasma Membrane Associations: A Continuous System of Plasma Membrane-Generating Transition Elements?

Many types of growing cells lack the familiar morphological pattern associated with the surface discharge of Golgi apparatus-derived secre-

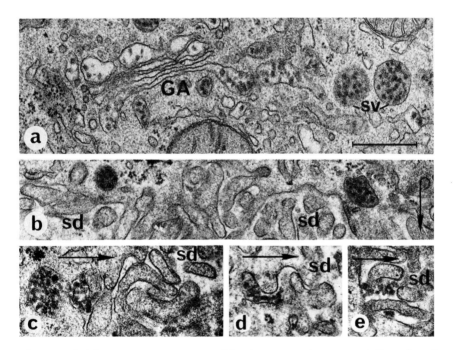

FIG. 9. Golgi apparatus–secretory vesicle–plasma membrane associations in rat liver. (a) Portion of a Golgi apparatus (*GA*) showing the dominant type of secretory vesicle (*sv*) which characterizes rat liver. In this particular fixation, the vesicles contained osmiophilic particles 300–600 Å in diameter in an electron-dense matrix. The particles are lipoproteins, precursors to the very-low-density circulating lipoproteins (Mahley *et al.*, 1970). (b) Similar secretory vesicles are found near the space of Disse (*sd*) at the cell surface. (c–e) Clusters of particles, apparently derived from secretory vesicles, appear to be secreted into the space of Disse (*sd*). Arrow indicates the location of the free surface of the cell. Bar = 0.5 μm.

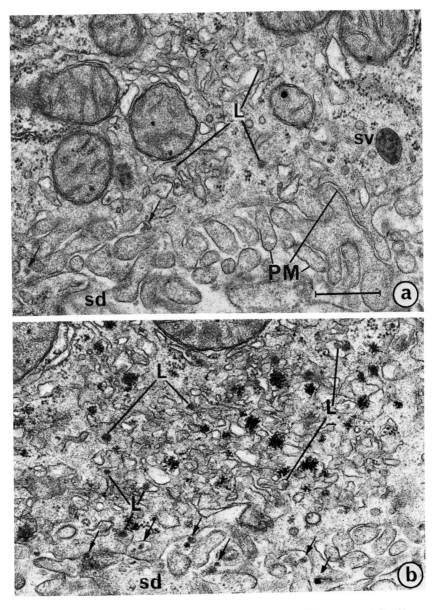

Fɪɢ. 10. Portions of the cytoplasm bordering the capillary space (*sd*). Shown is a secretory vesicle (*sv*) and the extensive system of endoplasmic reticulum characteristic of the cytoplasm near the cell surface. Continuities occur between rough-surfaced endoplasmic reticulum and smooth-surfaced endoplasmic reticulum and possibly between smooth-surfaced endoplasmic reticulum and the plasma mem-

tory vesicles. Additionally, data of Fig. 7 show two distinct phases of incorporation of labeled amino acids into proteins of the plasma membrane. The first phase is complete in about 20 minutes and lags only slightly behind incorporation into rough-surfaced endoplasmic reticulum (but reaches a maximum in advance of the Golgi apparatus). The second phase begins about 20 minutes after administration of label and parallels the discharge of a secretory product, such as serum albumin, into the circulation (plasma membrane corrected). Similar results were observed in experiments with onion stem, where the kinetics of leucine incorporation into the plasma membrane fraction showed two phases (Fig. 8): an initial incorporation phase paralleling incorporation into dictyosomes and a second phase which, by extrapolation, lags approximately 30 minutes behind the initial phase.

If we equate the second phase of incorporation of radioactive amino acids into plasma membrane protein with the fusion of secretory vesicles derived from the Golgi apparatus, how does one account for the initial phase of incorporation? Because of the extent of incorporation and their unique kinetic behavior, it is difficult to eliminate either phase on the basis of "microsomal contamination" (see also data of Ray et al., 1968).

A potential ultrastructural basis for interpreting the initial phase of incorporation of radioactive amino acids into the plasma membrane of rat liver is provided in Figs. 9 and 10. Reexamination of cytoplasmic regions adjacent to the capillary spaces revealed two systems of membranous components which appeared to be involved in the discharge of secretory products and which, at the same time, provide parallel mechanisms for transfer of membrane from rough-surfaced endoplasmic reticulum to the plasma membrane.

The first mechanism, already discussed, is provided by secretory vesicles. The predominant type of secretory vesicle produced by hepatocyte Golgi apparatus is illustrated in Fig. 9. The vesicles are characterized by osmiphilic particles which correspond to precursors of the very low density circulating lipoproteins. The particles are 300–600 Å in diameter and are contained in an electron-dense matrix. These vesicles originate

---

brane. Lipoprotein particles of slightly greater diameter (500–1000 Å) than those of the secretory vesicles occur singly or in small clusters within the cavities of the smooth endoplasmic reticulum (L). These particles lack the electron dense matrix characteristic of mature secretory vesicles and correspond to liposomes (cf. Hamilton, 1968). Single particles of the endoplasmic reticulum type are found near the plasma membrane and appear to be secreted directly into the extracellular space (arrows). These latter images are interpreted as resulting from the fusion of smooth-surfaced endoplasmic reticulum or vesicles derived from the smooth-surfaced endoplasmic reticulum directly with the plasma membrane. Bar = 0.5 μm.

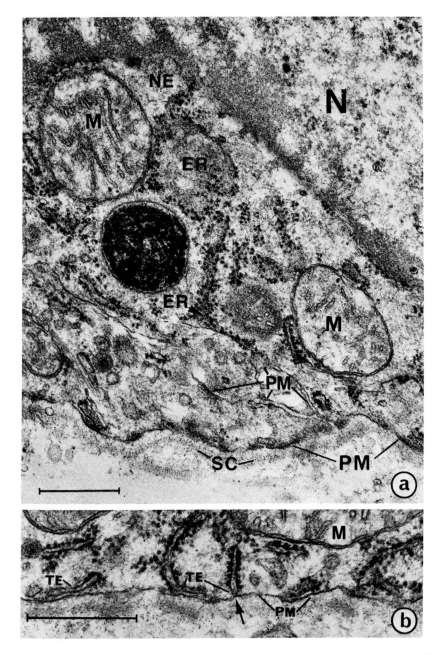

Fig. 11. Electron micrographs of portions of cells from the mammary gland of the rat. (a) An extension of the nuclear envelope (*NE*) continuous with rough-surfaced endoplasmic reticulum (*ER*) and extending to a region of association

at the Golgi apparatus (Fig. 9a), are found in significant numbers at or near the luminal surface of the cell (Figs. 9b–e), and appear to fuse with the plasma membrane to effect the discharge of the vesicle contents (see Hamilton, 1968).

The second mechanism, implicit in the observations of both Jones (1968) and Hamilton (1968) is provided by the extensive system of smooth-surfaced endoplasmic reticulum at the cell surface, directly continuous with rough-surfaced endoplasmic reticulum and also containing osmiophilic particles (Fig. 10). The particles of the smooth endoplasmic reticulum, however, are frequently larger than those of secretory vesicles (500–1000 Å), occur singly or in small clusters, and lack any surrounding electron-dense matrix (Fig. 10). These particle-containing profiles of smooth-surfaced endoplasmic reticulum are found in close proximity to the plasma membrane. Additionally, we observe particles resembling those of the smooth-surfaced endoplasmic reticulum in the extracellular space as well as images which appear to result from the coalesence of the particle-containing smooth endoplasmic reticulum with the plasma membrane (Fig. 10). It seems unrealistic for lipoprotein particles of the smooth-surfaced endoplasmic reticulum to migrate from near the cell surface back to the Golgi apparatus of the cell's interior and then back to the cell surface as secretory vesicles. A more realistic route would be transfer directly to the cell surface via these smooth-surfaced transition elements which could function, in effect, as Golgi apparatus equivalents. Close associations between endoplasmic reticulum and plasma membrane are observed frequently in other cell types (cf. Figs. 11 and 12) but, in the absence of a specific marker such as is provided by the lipoprotein particles in rat liver, functional continuity is more difficult to establish.

Additional insight into the question of Golgi apparatus equivalents has come from studies with fungi, many of which do not possess the stacks of cisternae that usually characterize the Golgi apparatus (Bracker, 1968; Girbardt, 1969; Morré et al., 1971d; Grove and Bracker, 1970),

---

with the plasma membrane (PM). It is not possible to trace luminal continuity from the cell exterior to the nucleoplasm in this micrograph, nor do we suggest that extensive or permanent luminal continuity between plasma membrane and endoplasmic reticulum are essential features of membrane flow or even normal cell function. As emphasized by Bracker and Grove (1971), luminal continuity need not be a requisite for transfer or exchange of membrane constituents. Contact may be sufficient. M = mitochondrion; SC = surface coat material; N = nucleus. Bar = 0.5 μm. (b) Specialized endoplasmic reticulum profiles of the cell surface which lack ribosomes at or near regions of close association (arrow) with the plasma membrane. A possibility we wish to consider is that these structures function as a form of transition element (TE) in membrane flow and differentiation. M = mitochondrion. Bar = 0.25 μm. From a study with Professor T. W. Keenan.

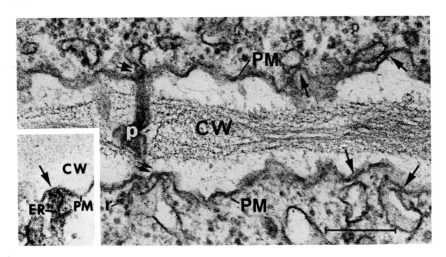

FIG. 12. Electron micrographs of plant cells (onion stem); cells of tissues similar to those used in experiments of Fig. 8. Regions of close association between endoplasmic reticulum and plasma membrane (double arrows) are most prevalent at intracellular connections or plasmodesmata (*p*) but are not restricted to these sites (single arrows). Inset shows a region of association between plasma membrane (*PM*) and a tubular segment of endoplasmic reticulum (*ER*) lacking ribosomes. Luminal continuity is not shown, but membrane contact or continuity remains as a possibility to be considered. *CW* = cell wall; *r* = ribosome. Bar = 0.25 μm. From a study with Professor C. A. Lembi.

yet contain secretory vesicles and exhibit surface growth by vesicular additions. Here, secretory vesicles are formed directly from short tubular or inflated, smooth-surfaced cisternae. These cisternae show functional continuity with rough-surfaced endoplasmic reticulum, serve as transition elements and function as equivalents of the Golgi apparatus (Morré *et al.*, 1971d).

The various possibilities for systems of transition elements involved in flow and differentiation of membranes from endoplasmic reticulum to the plasma membrane are diagrammed in Fig. 13. At least two pathways, operating in concert or independently, appear both necessary and suffi-

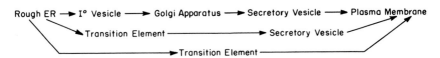

FIG. 13. Diagram summarizing the functional similarities and differences among Golgi apparatus (top), fungal transition elements (center) and direct endoplasmic reticulum(*ER*)-plasma membrane associations (bottom) as systems of plasma membrane-generating transition elements.

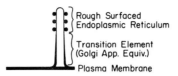

Fig. 14. Diagram of the hypothetical "minimal transition element." Included in the structural unit are rough-surfaced endoplasmic reticulum as the site of membrane synthesis, a smooth-surfaced transitional region as the site of membrane differentiation (Golgi apparatus equivalent), and a site of membrane transfer to the plasma membrane. Luminal continuity is not indicated. Although an essential feature of secretory processes involving products contained within the luminal space, luminal continuity is only one of several types of membrane associations illustrated by Bracker and Grove (1971) and is lacking from certain types of endoplasmic reticulum–mitochondrial outer membrane associations (Bracker and Grove, 1971; Morré et al., 1971c).

cient to explain the kinetic evidence for incorporation of amino acids into plasma membrane proteins (Figs. 7 and 8) and to account for the widely varying ultrastructural patterns associated with different types of growing cells. Thus, we visualize the origin of the plasma membrane as a multistep process beginning with initial synthesis of the membrane at the level of rough-surfaced endoplasmic reticulum followed by membrane flow and differentiation via a system of transition elements to the cell surface. In this scheme, the Golgi apparatus represents only one of several types of transition element which might function in this process. Although hypothetical at this stage, the minimum structural requirements of the type of transition element visualized are summarized in Fig. 14. Implicit in the concept are mechanisms whereby membrane constituents are sequentially added, deleted, altered, or rearranged to effect the necessary biochemical changes in the conversion of endoplasmic reticulum membrane to plasma membrane (Morré et al., 1971a,b,d).

IV. SUMMARY

Evidence is provided from plant and animal cells for the origin of the cell surface as the product of the integrated activities of the cell's internal (endomembrane) system. Proteins of the plasma membrane appear to be synthesized at the level of rough-surfaced endoplasmic reticulum and transferred to the cell surface via a system of discontinuous transition elements including the Golgi apparatus and derived vesicles or perhaps even more directly via a system of continuous transition elements which more directly link rough-surfaced endoplasmic reticulum with the plasma membrane. Pollen tubes of Easter lily provide an example of the origin of a cell surface where Golgi apparatus-derived secretory vesicles contribute both plasma membrane and matrix polysaccharides of the surface coat while deposition and organization of the microfibrillar phase of the coat occurs at the cell surface in a manner independent of secretory vesicle additions. We visualize the production of polysaccharide surface coats to result from the action of multiglycosyltransferase complexes associated generally with components of the endomembrane system, but subject to precise regulation to ensure proper sequential functioning.

ACKNOWLEDGMENTS

We thank Professors C. A. Lembi and T. W. Keenan, Purdue University for critically reading the manuscript and for many contributions to the experiments reported. We are grateful to Mr. Denis Bailey, Department of Biochemistry, University of Toronto and Dr. W. W. Franke, University of Freiburg, for use of unpublished information and to Professor Walter A. Rosen, New York State University at Buffalo for introducing us to the pollen tube growth system. Work supported in part by grants from the NSF GB 1084, 03044, 7078, and 2358. Purdue University AES Journal Paper No. 4477.

REFERENCES

ALBERSHEIM, P. (1965). Biogenesis of the cell wall. *In* "Plant Biochemistry" (J. Bonner and J. E. Varner, eds.), 2nd ed., pp. 298–321. Academic Press, New York.

ALVES, L. M., MIDDLETON, A. E., and MORRÉ, D. J. (1968). Localization of callose deposits in pollen tubes of *Lilium longiflorum* Thunb. by fluoresence microscopy. *Proc. Indiana Acad. Sci.* **77**, 144–147

BENEDETTI, E. L., and EMMELOT, P. (1968). Structure and function of plasma membranes isolated from liver. *In* "The Membranes" (A. Dalton and F. Haguenau, eds.), pp. 33–120. Academic Press, New York.

BENNETT, H. S. (1963). Morphological aspects of extracellular polysaccharides. *J. Histochem. Cytochem.* **11**, 15–23.

BONNETT, H. T., and NEWCOMB, E. H. (1966). Coated vesicles and other cytoplasmic components of growing root hairs of radish. *Protoplasma* **62**, 59–75.

BRACKER, C. E. (1968). The ultrastructure and development of sporangia in *Gilbertella persicaria*. *Mycologia* **60**, 1016–1017.

BRACKER, C. E., and GROVE, S. N. (1971). Continuity between cytoplasmic endomembrane and outer mitochondrial membranes in fungi. *Protoplasma* **73**, 15–34.

BREW, K. (1969). Secretion of α-lactalbumin into milk and its relevance to the organization and control of lactose synthetase. *Nature (London)* **222**, 671–672.

BREWBAKER, J. L., and KWACK, B. H. (1964). The calcium ion and substances influencing pollen growth. *In* "Pollen Physiology and Fertilization" (H. F. Linskens, ed.), pp. 143–151. North-Holland Publ., Amsterdam.

BRODBECK, U., and EBNER, K. E. (1966). Resolution of a soluble lactose synthetase into two protein components and solubilization of microsomal lactose synthetase. *J. Biol. Chem.* **241**, 762–764.

BROWN, R. M., FRANKE, W. W., KLEINIG, H., FALK, H., and SITTE, P. (1970). Scale formation in chrysophycean algae. I. Cellulosic and noncellulosic wall components made by the Golgi apparatus. *J. Cell Biol.* **45,** 246–271.

CURTIS, A. S. G. (1967). "The Cell Surface: Its Molecular Role in Morphogenesis." Academic Press, New York.

DASHEK, W. V., and ROSEN, W. G. (1966). Electron microscopical localization of chemical components in the growth zone of lily pollen tubes. *Protoplasma* **61,** 192–204.

DAVIS, B. D., and WARREN, L., eds. (1967). "The Specificity of Cell Surfaces." Prentice-Hall, Englewood Cliffs, New Jersey.

DOWBEN, R. M., ed. (1969). "Biological Membranes." Little, Brown, Boston, Massachusetts.

FLEISCHER, B., and FLEISCHER, S. (1970). Preparation and characterization of Golgi membranes from rat liver. *Biochim. Biophys. Acta* **219,** 301–319.

FRANKE, W. W., MORRÉ, D. J., DEUMLING, B., CHEETHAM, R. D., KARTENBECK, J., JARASCH, E.-D., and ZENTGRAS, H.-W. (1971). Synthesis and turnover of membrane proteins in rat liver; an examination of the membrane flow hypothesis. *Z. Naturforsch.* **26b,** 1031–1039.

GIRBARDT, M. (1969). Die Ultrastruktur der Apikalregion von Pilzhypheo. *Protoplasma* **67,** 413–441.

GROVE, S. N., and BRACKER, C. E. (1970). Protoplasmic organization among fungi: Vesicles and spitzenkörper. *J. Bacteriol.* **104,** 989–1009.

GROVE, S. N., BRACKER, C. E., and MORRÉ, D. J. (1968). Cytomembrane differentiation in the endoplasmic reticulum-Golgi apparatus-vesicle complex. *Science* **161,** 171–173.

GROVE, S. N., BRACKER, C. E., and MORRÉ, D. J. (1970). An ultrastructural basis for hyphal tip growth in *Pythium ultimum. Amer. J. Bot.* **57,** 245–266.

HAMILTON, R. L. (1968). Ultrastructural aspects of hepatic lipoprotein synthesis and secretion. *Proc. Deuel Conf. Lipids, 1968* pp. 3–31.

HARRIS, P. J., and NORTHCOTE, D. H. (1971). Polysaccharide formation in plant Golgi bodies. *Biochim. Biophys. Acta* **237,** 56–64.

HASSID, Z. (1969). Biosynthesis of oligosaccharides and polysaccharides in plants. *Science* **165,** 137–144.

HOUSE, P. D. R., and WIEDEMANN, M. J. (1970). Characterization of an (I^{125}) insulin binding plasma membrane from rat liver. *Biochem. Biophys. Res. Commun.* **41,** 541–548.

JENSEN, W. A., and FISHER, D. B. (1970). Cotton embryogenesis: The pollen tube in the stigma and style. *Protoplasma* **69,** 215–235.

JONES, A. L. (1968). Some electron microscopic observations on lipoprotein and chylomicron metabolism in perfused liver. *Proc. Deuel Conf. Lipids, 1968* pp. 33–40.

JONES, B. M., and MORRISON, G. A. (1969). A molecular basis for indiscriminate and selective cell adhesion. *J. Cell Sci.* **4,** 799–813.

KEENAN, T. W., MORRÉ, D. J., and CHEETHAM, R. D. (1970). Lactose synthetase in a Golgi apparatus fraction from rat mammary gland. *Nature (London)* **228,** 1105–1106.

KEENAN, T. W., HUANG, C.-M., and MORRÉ, D. J. (1973). Membrane of mammary gland. II. Lipid composition of Golgi apparatus from rat mammary gland. *J. Dairy Sci.* **55,** 51–57.

LARSON, D. (1965). Fine structural changes in the cytoplasm of germinating pollen. *Amer. J. Bot.* **52,** 139–154.

MAHLEY, R. W., BERSOT, T. P., and LEQUIRE, V. S. (1970). Identity of very low density lipoproteins of plasma and liver Golgi apparatus. *Science* **168**, 380–382.

MANTON, I. (1966). Observations on scale production in *Prymnesium parvum*. *J. Cell Sci.* **1**, 375–380.

MANTON, I. (1967). Further observations on scale formation in *Chrysochromulina chiton*. *J. Cell Sci.* **2**, 411–418.

MASCARENHAS, J. P. (1970). A new intermediate in plant cell wall synthesis. *Biochem. Biophys. Res. Commun.* **41**, 142–149.

MASON, L. A., ed. (1968). "Biological Properties of the Mammalian Surface Membrane." Wistar Inst. Press, Philadelphia, Pennsylvania.

MOLLENHAUER, H. H. (1971). Fragmentation of mature dictyosome cisternae. *J. Cell Biol.* **49**, 212–214.

MOLLENHAUER, H. H., and MORRÉ, D. J. (1966). Golgi apparatus and plant secretion. *Annu. Rev. Plant Physiol.* **17**, 27–46.

MOLLENHAUER, H. H., and WHALEY, W. G. (1963). An observation on the functioning of the Golgi apparatus. *J. Cell Biol.* **17**, 222–225.

MORRÉ, D. J. (1970). *In vivo* incorporation of radioactive metabolites by dictyosomes and other cell fractions of onion stem. *Plant Physiol.* **45**, 791–799.

MORRÉ, D. J., FRANKE, W. W., DEUMLING, B., NYQUIST, S. E., and OVTRACHT, L. (1971a). Golgi apparatus function in membrane flow and differentiation: Origin of plasma membrane from endoplasmic reticulum. *In* "Biomembranes" (L. A. Meson, ed.), Vol. 2, pp. 95–104. Plenum Press, New York.

MORRÉ, D. J., KEENAN, T. W., and MOLLENHAUER, H. H. (1971b). Golgi apparatus function in membrane transformations and product compartmentalization: Studies with cell fractions from rat liver. *In* "Advances in Cytopharmacology" (S. Clementi and B. Ceccarelli, eds.) Vol. 1, pp. 159–182. Raven Press, New York.

MORRÉ, D. J., MERRITT, W. D., and LEMBI, C. A. (1971c). Connections between mitochondria and endoplasmic reticulum in rat liver and onion stem. *Protoplasma* **73**, 43–49.

MORRÉ, D. J., MOLLENHAUER, H. H., and BRACKER, C. E. (1971d). Origin and continuity of the Golgi apparatus. *In* "Results and Problems in Cell Differentiation" (T. Reinert and H. Ursprung, eds.), Vol. 2, pp. 82–126. Springer-Verlag, Berlin and New York.

NORTHCOTE, D. H. (1968). Structure and function of plant cell membranes. *Brit. Med. Bull.* **24**, 107–112.

NORTHCOTE, D. H. (1971). The Golgi apparatus. *Endeavour* **30**, 26–33.

ORDIN, L., and HALL, M. A. (1968). Cellulose synthesis in higher plants from UDP-glucose. *Plant Physiol.* **43**, 473–476.

PICKETT-HEAPS, J. D. (1968). Further ultrastructural observations of polysaccharide localization in plant cells. *J. Cell Sci.* **3**, 55–64.

PRESTON, R. D. (1959). Cellulose-protein complexes in plant cell walls. *In* "Macromolecular Complexes" (M. V. Edds, ed.), pp. 229–253. Ronald Press, New York.

RAMBOURG, A., NEUTRA, M., and LEBLOND, C. P. (1966). Presence of a "cell coat" rich in carbohydrate at the surface of cells in the rat. *Anat. Rec.* **154**, 41–72.

RAY, T. K., LIEBERMAN, I., and LANSING, A. I. (1968). Synthesis of the plasma membrane of the liver cell. *Biochem. Biophys. Res. Commun.* **31**, 54–58.

REVEL, J. P., and ITO, S. (1966). The surface component of cells. *In* "The Specificity of Cell Surfaces" (B. Davis and L. Warren, eds.), pp. 211–234. Prentice-Hall, Englewood Cliffs, New Jersey.

ROSEMAN, S. (1970). The synthesis of complex carbohydrates by multiglycosyltransferase systems and their potential function in intercellular adhesion. *Chem. Phys. Lipids* **5**, 270–297.

ROSEN, W. G. (1968). Ultrastructure and physiology of pollen. *Annu. Rev. Plant Physiol.* **19**, 435–462.

ROSEN, W. G., and GAWLICK, S. R. (1966). Fine structure of lily pollen tubes following various fixation and staining procedures. *Protoplasma* **61**, 181–191.

ROSEN, W. G., GAWLICK, S. R., DASHEK, W. V., and SIEGESMUND, K. A. (1964). Fine structure and cytochemistry of Lilium pollen tubes. *Amer. J. Bot.* **51**, 61–71.

SASSEN, M. M. A. (1964). Fine structure of petunia pollen grain and pollen tube. *Acta Bot. Neer.* **13**, 175–181.

TURKINGTON, R. W. (1970). Stimulation of RNA synthesis in isolated mammary cells by insulin and prolactin bound to Sepharose. *Biochem. Biophys. Res. Commun.* **41**, 1362–1367.

VAN DER WOUDE, W. J., and MORRÉ, D. J. (1968). Endoplasmic reticulum-dictyosome-secretory vesicle associations in pollen tubes of *Lilium longiflorum* Thumb. *Proc. Indiana Acad. Sci.* **77**, 164–170.

VAN DER WOUDE, W. J., MORRÉ, D. J., and BRACKER, C. E. (1969). A role for secretory vesicles in polysaccharide biosynthesis. *Abst. Int. Bot. Congr., 11th, 1969,* p. 226.

VAN DER WOUDE, W. J., MORRÉ, D. J., and BRACKER, C. E. (1971). Isolation and characterization of secretory vesicles in germinated pollen of *Lilium longiflorum. J. Cell Sci.* **8**, 331–351.

VILLIMEZ, C. L., MCNABB, J. M., and ALBERSHEIM, P. (1968). Formation of plant cell wall polysaccharides. *Nature (London)* **218**, 878–880.

YOUNG, L. C. T., STANLEY, R. G., and LOEWUS, F. A. (1966). Myo-inositol-2-t incorporation by germinating pollen. *Nature (London)* **209**, 530–531.

YUNGHANS, W., KEENAN, T. W., and MORRÉ, D. J. (1970). Isolation of a Golgi apparatus-rich cell fraction from rat liver. III. Lipid and protein composition. *Exp. Mol. Pathol.* **12**, 36–45.

The Role of Hydroxyproline-Rich Proteins in the Extracellular Matrix of Plants

Derek T. A. Lamport

*MSU/AEC Plant Research Laboratory, Michigan State University,
East Lansing, Michigan*

I. INTRODUCTION

Honesty and the desire for self-preservation prompt me to begin with a negative. Despite a recent editorial in *Nature (London), New Biology,* we still do not know the role of hydroxyproline-rich glycoproteins in plants. The name "extensin" is still provisional, and only suggests the possibility that this protein plays a role in cell extension. This presentation is an attempt to discuss some of the more likely roles for a protein in the extracellular matrix of plants. I use the term extracellular matrix as a general term for several reasons: It is more inclusive than cell wall because it applies to structures which some purists might not regard as cell walls (for example, the extracellular materials secreted by some algae). I think of a cell wall as a type of extracellular matrix which is highly organized and which generally retains its form on isolation. The term extracellular matrix also describes the protein-polysaccharide complexes secreted by animal cells (if Robert Hooke had sectioned cartilage instead of cork perhaps chondrocytes would have cell walls?). The extracellular matrices of plants and animals have similarities which are so striking that at least one observer (Aaronson, 1970) boldly suggests that animal skeletons evolved from plant cell walls.

But before grappling with that type of speculation, I shall review the evidence for the very existence of a hydroxyproline-rich component outside the plasma membrane, especially because there still exists a school of skepticism on this fundamental point (Steward *et al.*, 1967). Then we shall consider the chemistry of this macromolecule. Such an approach may help to answer some leading questions. For example, is the peptide component of "glycoprotein," protein per se? Is it covalently attached to wall polysaccharide? Does it increase the tensile strength of cell walls? Is it ubiquitous? Is it a requirement (e.g., a primer) for polysaccharide synthesis or transport? Do hydroxyproline-rich glycoproteins occur exclusively in the cell wall or are some specifically cytoplasmic? Is it localized in discrete areas of the wall, or randomly or uniformly distributed? Is it enzymatic? Is it part of a self-assembly system? Is it involved in disease resistance or frost hardiness? Is it involved in cell extension?

II. LOCALIZATION OF HYDROXYPROLINE

All primary cell walls isolated from higher plants contain firmly bound hydroxyproline ranging in amount from 0.05 to 2.7% of the cell wall on a dry weight basis. Arguments arise only when one deals with the lower end of the scale, for at this level cytoplasmic contamination is a possibility. For example, the oat coleoptile wall is about 0.05% hydroxyproline, and this represents only 50–60% of the total cellular hydroxyproline. Generally more than 90% of the cellular hydroxyproline is associated with the cell wall fraction. These comments apply in a limited way to the lower plants, especially among the red and brown algae (Rhodophyceae and Phaeophyceae), where hydroxyproline is absent or is present only in vanishingly small quantities (Gotelli and Cleland, 1968). On the other hand, the green algae, the group from which land plants arose, contain (with rare exceptions) hydroxyproline in the extracellular matrix. Some of these algae, such as *Volvox*, have no clearly defined wall but consist of cells embedded in a mucilaginous hydroxyproline-rich matrix. Volvox provides the simplest demonstration of extracellular hydroxyproline because proteolytic treatment solubilizes the hydroxyproline-rich matrix and releases intact cells (D. T. A. Lamport and G. Kochert, unpublished results, 1971).

III. CHEMISTRY OF HYDROXYPROLINE-RICH GLYCOPROTEINS

A. *Glycopeptides and Hydroxyproline Arabinosides*

Unfortunately walls isolated from higher plants do not degrade as easily as *Volvox* extracellular matrix. For example, walls isolated from cultures of tomato or sycamore-maple are fairly resistant to proteolytic attack. The resistance might be due to steric hindrance from wall polysaccharide and/or a protease-resistant amino acid sequence. We shall see that both factors are important possibly also as a deterrent to some investigators.

However, some enzymes, such as crude cellulase and pronase, do release hydroxyproline-rich glycopeptides from cell walls. Some walls are much more resistant than others. Naturally we work with the susceptible ones confident that this will ultimately show us how to degrade the resistant walls. The crude cellulase from *Aspergillus niger* exhibits proteolytic activity besides polysaccharidase activity. It released up to 75% of the hydroxyproline from some walls (e.g., tomato, tobacco) and as little as 5% from walls such as potato. Fractionation of the released materials by gel filtration and ion exchange chromatography gave us several glycopeptides, each of which had in common, arabinose, galactose, hydroxyproline, and serine (Lamport, 1969).

Attachment of the carbohydrate must be through a hydroxyproline or a serine residue, or possibly both. The glycopeptides are stable in weak alkali (0.5 N KOH 18 hours, 4°C) but are hydrolyzed by strong alkali (0.22 M Ba(OH)$_2$ 6 hours, 100°C) yielding free amino acids plus hydroxyproline-O-arabinosides which, unlike peptidylhydroxyproline, react with Ehrlich's reagent after oxidation.

The arabinosides were identified, after separation by electrophoresis or cation exchange chromatography (Fig. 1) by showing that both the hydroxyproline secondary amino and carboxyl groups were free (neutral on pH 6.5 electrophoresis), leaving the hydroxyl group as the most likely point of substitution. The arabinosides do not reduce alkaline silver nitrate, indicating the arabinose reducing end is blocked, or, to put it another way, arabinose C-1 is linked glycosidically to C-4 of hydroxyproline. Arabinosides are well known to be acid labile; hydroxyproline arabinosides are no exception. For example pH 1 treatment for 1 hour at 100° completely cleaves the arabinosyl hydroxyproline linkage. This observation is the key to the chemistry that unlocks the amino acid sequence because it allowed us to test the prediction that *removal of poly-*

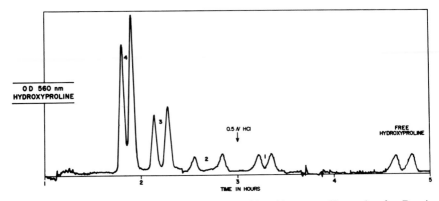

FIG. 1. Separation of hydroxyproline arabinosides on Chromobeads B. A Ba(OH)₂ hydrolyzate (0.22 M, 6 hours, 100°C) of sycamore-maple cell walls was neutralized with H_2SO_4, centrifuged, and then applied to a 75 × 0.6 cm column of Technicon Chromobeads B and eluted with a pH gradient; the column eluate was monitored continuously for non-peptide-bound hydroxyproline as described by Lamport and Miller (1971). The numbered double peaks (1–4) correspond to the racemic mixtures of hydroxyproline tetraarabinoside, triarabinoside, diarabinoside, and monoarabinoside, respectively.

saccharide would expose the polypeptide backbone to proteolytic attack. But before discussing the amino acid sequences obtained, I shall summarize what is known about the structure and biosynthesis of hydroxyproline arabinosides and their distribution in the plant kingdom. I shall deal in the next section with their possible turnover during growth.

The predominant hydroxyproline arabinoside released by alkaline hydrolysis of cell walls from higher plants contains four arabinose residues.

We were able to isolate enough of this hydroxyproline tetraarabinoside (from tomato) to obtain an elemental analysis consistent with its formula and an optical rotation (after acid hydrolysis) characteristic of L-arabinose. Dr. Arthur Karr in my laboratory continued with the structural analysis. Tritylation indicated the presence of arabinofuranose. Methylation of tetraarabinosylhydroxyproline followed by hydrolysis, reduction and acetylation gave a mixture of methylated alditol acetates which were separated by gas chromatography. The data obtained allow us to put forward a tentative structure for the tetraarabinoside as consisting of a linear chain of sugar residues in which the linkage sequence is ₍f₎A 1 ⟶ 3 ₍f₎A 1 ⟶ 2 ₍f₎A 1 ⟶ 3 ₍f₎A 1 ⟶ 4 Hyp (Fig. 2). The anomeric configuration around C-1 remains to be determined, although the β-L configuration seems likely in view of the resistance of the

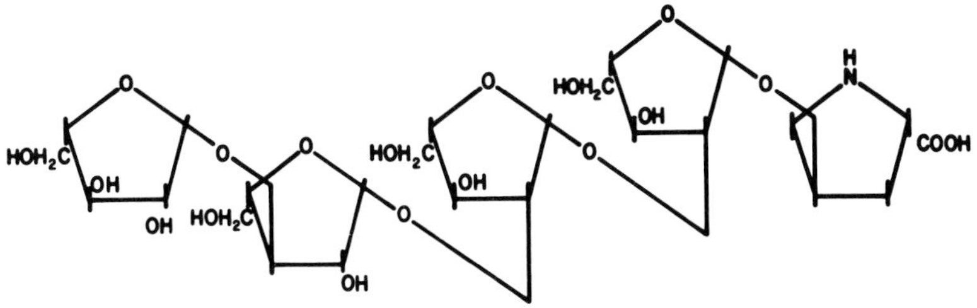

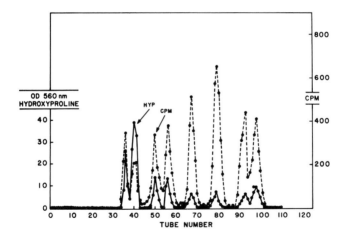

FIG. 2. Hydroxy-L-proline tetra-L-arabinofuranoside. This is the tentative structure for the hydroxyproline tetraarabinoside isolated from sycamore-maple cell walls.

tetraarabinoside to emulsin, a preparation reported to exhibit α-L-arabinofuranosidase activity.

Dr. Karr has also isolated a cell-free system (from cultured sycamore-maple cells) which catalyzes the transfer of arabinose from UDP-L-arabinose to peptidylhydroxyproline (Karr, 1972). The usual series of hydroxyproline arabinosides is produced (Fig. 3) although the mono-

FIG. 3. Biosynthesis of hydroxyproline arabinosides in a cell-free system. Sycamore-maple cells were sonicated for 60 seconds in 50 mM Tris, pH 6.9, containing 400 mM sucrose, 1% bovine serum albumin, and 4 mM sodium metabisulfite. The homogenate was centrifuged 1000 g for 15 minutes and the pellet was discarded. The supernatant was centrifuged at 37,000 g for 60 minutes and the pellet (= enzyme preparation) was resuspended in 50 mM sodium acetate pH 6.5 containing 400 mM sucrose and 1% bovine serum albumin. Reaction mixtures were prepared by mixing 200 μl of enzyme preparation with UDP-L-[^{14}C]arabinose. The product was mixed with carrier and subjected to alkaline hydrolysis and chromatography as described in Fig. 1 legend.

and diarabinosides predominate, suggesting that the short arabinoside chains are built up by the addition of single residues to hydroxyproline rather than the transfer of complete oligosaccharides. We also found that the addition of high molecular weight *tomato* tryptides to the sycamore-maple homogenizing medium enhanced the *in vitro* incorporation of L-[^{14}C]arabinose (from UDP-L-arabinose) into the hydroxyproline arabinosides. This is not unexpected and merely indicates that arabinose incorporation in the cell-free system is limited by the protein acting as a primer. We should therefore consider another possible (but not necessarily exclusive) function for the protein, that is, as a matrix for assembly, transport, and transfer of wall polysaccharides. In fact the particulate system which catalyzes the arabinoside biosynthesis is similar to Villemez' system for polysaccharide biosynthesis, which also appears to require glycoprotein addition for sustained activity (Villemez, 1970).

The occurrence of hydroxyproline arabinosides might be an interesting triviality restricted to a few species, such as sycamore-maple and tomato. But even a limited survey of the plant kingdom (Lamport and Miller, 1971) shows that wherever cell wall hydroxyproline occurs it is arabinosylated to a greater or lesser extent (Table 1). This common structural feature leads me to speculate that first-formed, so-called primary walls, from green algae to the higher plants are distinctly homologous, built up from a hydroxyproline-rich protein–glycan network analogous to the bacterial peptidoglycan. Cellulose, virtually synonymous with cell wall,

TABLE 1

HYDROXYPROLINE ARABINOSIDE SURVEY: SUMMARY[a]

Source	Number of arabinose residues in predominant arabinoside
Algae[b]	2
Mosses	2
Liverworts	3
Ferns	3 (+4)
Horsetails	3 (+4)
Gymnosperms	3 (+4)
Monocots	3 (+4)
Angiosperms	
Dicots	4

[a] See Lamport and Miller (1971) for complete data.

[b] In *Chlamydomonas* and *Volvox* hetero-oligosaccharides containing galactose, glucose, and arabinose may also be attached to hydroxyproline.

is in fact absent from green algae such as *Codium*, which nevertheless contain wall-bound hydroxyproline (Thompson and Preston, 1967).

B. Peptides, Conformation, and Polysaccharide Involvement

Evidence accumulated over the last ten years (Lamport, 1965, 1970) is consistent with the idea that cell wall hydroxyproline is a protein component, implying that its synthesis involves specific mRNA and ribosomal synthesis. However, a specific mRNA and *in vitro* synthesis of this hydroxyproline-rich protein remain to be demonstrated.

Presumably doubts about the authenticity of a protein would be laid to rest if its amino acid sequence were known. But, as stated earlier, the fact that wall protein appears to have "built-in braces,"[1] has until recently foiled attempts of investigators (unpublished) to determine amino acid sequences. However, the discovery of the hydroxyproline-*O*-arabinosyl linkage and the realization of its extreme acid lability allowed us to test the prediction that removal of polysaccharide would expose the polypeptide backbone to proteolytic attack.

Thus trypsin released about 60% of the hydroxyproline from acid-stripped (pH 1, 1 hour, 100°C) walls of tomato cells grown in suspension culture; that is, approximately 10 mg of hydroxyproline was released from 1 gram of cell walls (dry weight basis).

Gel filtration of these tryptides yielded high and low molecular weight fractions. The low molecular weight fractions were resolved into several subfractions by chromatography on an Aminex AG 50 W $\times$ 2 cation exchange resin (Fig. 4) and further purified by pH 1.9/6.5 electrophoresis or chromatography on a more highly cross-linked resin (Aminex A5).

The compositions of the tryptic peptides (Table 2) are strikingly similar to the amino acid compositions of some of the glycopeptides previously isolated. Again the NH_2-terminus is serine and tryptic cleavage implicates lysine as C-terminal. Thus lysyl-serine, a linkage reported to show slow tryptic cleavage, occurs with high frequency. Lysyl hydroxyproline would probably be resistant to trypsin (cf. McBride and Harrington, 1967), and this might explain why acid-stripped cell walls of different species are *not* equally susceptible to tryptic attack. The small number of tryptic peptides means that there is probably only one hydroxyproline-rich protein in the wall. The approximate subunit size of this protein can be calculated from the analytical data which show that three peptide sequences (i.e., peptides 6–9) totaling 25 residues account for about 10% of the wall-bound hydroxyproline. On this basis, 100% of

---

[1] A term used by the late Dr. G. D. Greville when describing the poly-D-glutamic acid of *Bacillus subtilis*.

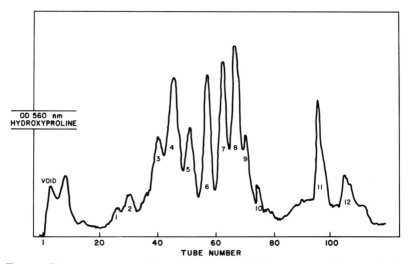

FIG. 4. Chromatography of low molecular weight tomato tryptides on Aminex AG 50W × 2 200–325 mesh. Tryptides containing 20–25 mg of hydroxyproline were loaded onto a 75 × 0.6 cm column equilibrated with pH 2.7 pyridine–acetic buffer, and then eluted with a pH gradient generated in a 9-chambered vessel containing 560 ml of pH 2.7, 280 ml of pH 3.1, and 140 ml of pH 5.0 pyridine–acetic buffers. Fractions were collected and analyzed for peptide-bound hydroxyproline on the automated hydroxyproline analyzer as described by Lamport and Miller (1971).

the wall-bound hydroxyproline totals 250 residues provided that (a) the 10% we looked at show average compositions, (b) these three peptide sequences were obtained in 100% yield (e.g., 50% yield would mean that they represent 20% of the wall protein or a total of $25 \times 5 = 125$ residues).

The sequencing of these peptides poses unusual problems: their structure makes further degradation by enzymatic attack less than promising while the repetitive hydroxyproline residues rule out qualitative methods of sequencing such as the dansyl/Edman. Fortunately, some of these peptides are small enough to be degraded by the classical method of partial acid hydrolysis, for example, by treatment with 6 N HCl for 10 minutes at 100°. Under these conditions cleavage occurs predominantly at the N-serine peptide bond. The products of partial acid hydrolysis were separated by paper electrophoresis, but more recently by column chromatography.

Results obtained so far by these methods show (Fig. 5) that all the tomato tryptides examined contain a common pentapeptide sequence: Ser Hyp Hyp Hyp Hyp. Partial acid hydrolysis of tomato tryptide 4 gave the series: Ser Hyp, Ser Hyp_2, Ser Hyp_3, and Ser Hyp_4, all with NH_2-terminal serine. These peptides have characteristic electrophoretic

TABLE 2

COMPOSITIONS OF TOMATO TRYPTIDES AND GLYCOPEPTIDES

Peptide	Empirical formula	NH_2 Terminus
Tryptide[a]		
3 4 5	Hyp_9 Ser_3 Tyr "U"[c] Lys	Ser
6 7	Hyp_5 Ser Lys	Ser
8	Hyp_6 Ser Thr Val Tyr Lys	Ser
9	Hyp_4 Ser Lys	Ser
11 12	Hyp_4 Ser Val "U" Lys_2	Ser
Glycopeptide[b]		
1	Ara_{25} Gal_6 Hyp_{10} Ser_3 Tyr	Ser
2	Ara_{14} Gal_3 Hyp_{10} Ser_3 Thr Val Lys_2	Ser
3	Ara_{20} Gal_4 Hyp_9 Ser_3 Tyr Lys	Ser
4	Ara_{18} Gal_4 Hyp_9 Ser_3 Tyr	Ser
5	Ara_{16} Gal_2 Hyp_9 Ser_3 Val Tyr Lys_3	Ser

[a] Number corresponds to those in Fig. 4.

[b] Taken from Lamport (1969).

[c] "U" refers to an unknown amino acid with properties somewhat similar to those of tyrosine.

(Fig. 6), and chromatographic mobilities and allowed us to confirm unequivocally the identification of the pentapeptide Ser Hyp_4 in the other tryptides. During the preparation of this paper we have also been able to isolate Ser Hyp_4 from sycamore-maple tryptides. The phylogenetic separation of tomato and sycamore-maple is sufficiently wide for me to conclude that the pentapeptide Ser Hyp_4 is of fundamental importance in higher plants.

In some respects the sequences of these tomato tryptides are distinctly uncollagen-like; glycine and the repeating tripeptide Gly Pro X are absent. On the other hand, both collagen and the tomato wall protein may possess the polyproline II conformation consisting of trans peptide bonds in a left-handed helix of approximately three residues per turn and 9.4 Å pitch. Despite the fact that the conformation of peptides rich in proline and hydroxyproline is fairly well worked out even to the extent of molecular orbital calculations (Maigret *et al.*, 1970), one cannot yet assume that sequences such as Ser Hyp_4 really do exist in a conformation similar to that of polyproline II. For example, is it purely fortuitous that we observe four, but never more than four, contiguous hydroxyproline residues? Or can it be related to the observation that in a series of proline

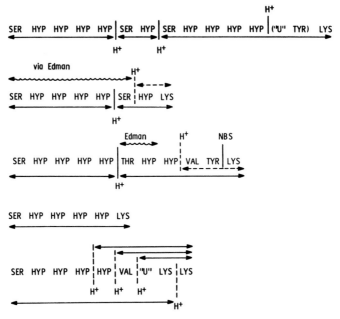

Fig. 5. Amino acid sequences of tomato tryptides. Horizontal arrows indicate peptides obtained after partial acid hydrolysis of the parent peptide. Vertical lines indicate cleavage points, a good yield being indicated by a continuous line, a low yield by a broken line. A wavy line indicated those residues sequenced by *subtractive* Edman degradation. NBS indicates that *N*-bromosuccinimide oxidation cleaved a tyrosyl bond. H⁺ indicates that partial acid hydrolysis (6 *N* HCl, 10 minutes, 100°C) was used to cleave peptide bonds. Note that peptides 6 and 7 are distinct electrophoretically and chromatographically, yet have identical sequences.

oligopeptides "... appearance of the helical conformation commences at the tetramer. When the number of residues is five or greater, the conformation of the helical structure of poly-L-proline II seems to be completed" (Okabayashi *et al.*, 1968)? Does the serine prevent this type of helix formation?

In collagen the strict occurrence of glycine at every third residue is a necessary steric feature because the absence of a side chain allows alignment of three peptide helices into the triple collagen helix. Obviously a highly glycosylated wall protein could never interact in this way; and from this point of view glycine residues are unnecessary. On the other hand, polysaccharide attached to the peptide will be oriented in directions determined by the peptide conformation. On this basis the usual descriptions of primary cell walls, as consisting of cellulose microfibrils in an

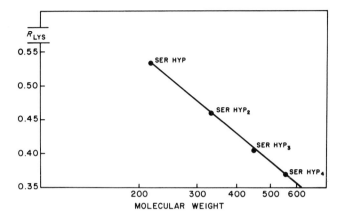

FIG. 6. Seryl hydroxyproline peptides. Electrophoretic mobilities on paper at pH 1.9 relative to lysine are plotted against $\log_{10}$ of the molecular weight.

amorphous hemicellulose matrix, are questionable. But what is the evidence for attachment of polysaccharide other than the arabinosides, to wall protein? Certainly the presence of small amounts of galactose in the glycopeptides obtained by enzymatic degradation of tomato cell walls (Lamport, 1969) supports the idea. Other evidence for polysaccharide per se attachment to wall protein is gathering slowly (Boundy et al., 1965, 1967; Jennings and Watt, 1967; Cleland, 1968; Chrispeels, 1969; Pusztai and Watt, 1969; Lamport, 1970). This evidence is in the form of polysaccharide–hydroxyproline–protein complexes isolated from both whole tissues (pericarp) and cytoplasmic cell fractions by extraction with trichloroacetic acid. These protein–polysaccharide complexes are soluble in trichloroacetic acid by virtue of the high proportion of polysaccharide. The polysaccharide component is predominantly arabinose and galactose together with other sugars. Unfortunately, in most instances little more is known of these complexes than their amino acid and sugar compositions, and often not even that. Some authors (e.g., Cleland, 1968) have suggested that, because these complexes occur in the cytoplasmic fraction and do not appear to enter the wall during [^{14}C]proline pulse-chase experiments, these complexes are not wall protein. This may or may not be correct. For example, complexes isolated in our laboratory yield hydroxyproline arabinosides on alkaline hydrolysis, and the amino acid composition is, apart from a high value for alanine, similar to that of wall protein (Lamport, 1970). Other investigators (Pusztai and Watt, 1969) fractionated these TCA-soluble complexes and concluded from their data that the amino acid compositions were markedly different and therefore indicated several hydroxyproline-rich proteins. But if one subtracts the ammonia "residues" and recalculates their data expressing each amino acid as a percentage of the total number of residues, one of their main fractions shows an amino acid composition remarkably similar to that of a typical analysis for tomato wall protein (Table 3). The problem

TABLE 3
AMINO ACID COMPOSITIONS COMPARED[a]

Residue	Tomato cell wall[b]	Tomato TCA-soluble[b]	Tomato β-glucosidase[c] HMW	Tomato β-glucosidase[c] LMW	Vicia faba TCA-soluble[d]	Volvox extra-cellular matrix[e]	Horse-radish perox-idase[f]
Hyp	30	30	30	30	30	30	30
Pro	8	ND	ND	ND	5	5.5	ND
Asp	8	5	35	48	8	12.5	27
Thr	6	12	13.5	23.5	7	10.5	16
Ser	15	19	44	52	13	11	32
Glu	9	6	7.5	27	8	5	23
Gly	8	9	37.5	50	8	25	25
Ala	7	26	17	31	9	11	25
Val	8	7	16	28	6	8	31
½-Cys	5	4	0	0	0	0	5
Met	1	1	0	0	1	1	0
Ile	5	2	13	19	2	3	10
Leu	9	4	18	32	3	6	18
Tyr	3	0.5	10	0	2	3	1
Phe	3	0.5	7.5	14.5	2	2	6
Lys	11	4	19	30	7	4	12
His	2	0.5	4.5	8	1	1	5
Arg	4	1	4.5	7	4	4	8
Total residues:	142	132.5	277	400	116	142.5	274

[a] The data were obtained from amino acid analyses which have been normalized to 30 residues of hydroxyproline in view of the calculation (see text) which indicates the fundamental wall protein subunit to be in the range of 125–250 residues.

[b] Taken from Lamport (1970).

[c] Taken from J. D. Ross (unpublished data, 1971). The enzyme was obtained by 0.5 M NaCl elution of tomato cell walls (from suspension cultures) followed by ammonium sulfate fractionation and gel filtration on Sephadex G-200. HMW refers to enzymatic activity which voided the column. LMW refers to the enzymatic activity which was retarded by the column.

[d] Data recalculated from Pusztai and Watt (1969): Table 5, "material soluble in the aqueous phase on phenol-borate partitioning."

[e] D. T. A. Lamport and G. Kochert (unpublished results, 1971). Analysis of Volvox matrix after release by proteolytic digestion and ethanol precipitation.

[f] E. Liu (unpublished data, 1971). An anodic isozyme of horseradish preoxidase.

will be solved if the cytoplasmic and wall hydroxyproline-rich proteins turn out to contain the same sequence.

At the moment then, we do not know how this polysaccharide is attached to a hydroxyproline-rich protein. An obvious possibility is attachment of polysaccharide chains via the arabinoside. (What else can the arabinosides do?)

We could then view the polypeptide backbone as providing cross-links for polysaccharide strands. For example, a short sequence of protein could cross-link a relatively large amount of polysaccharide:

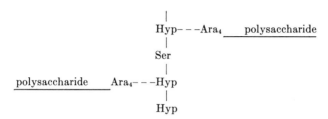

This model raises the problem of accounting for the alkaline lability of the (galactosyl-arabinose?) linkage, and because such weak linkages are rare in polysaccharides one is tempted to pile speculation on speculation by suggesting that if this linkage does indeed exist, its alkaline lability reflects an essential chemical property conservatively retained from the green algae to the flowering plants. This comparative type of argument may help us select for further scrutiny, those linkages of the (admittedly incomplete) extensin network that are enzymatically labile and under cellular control. The arabinoside linkages themselves have been scrutinized, but with negative results. If any of these linkages were broken we expected the arabinoside profile to reflect this at different stages of growth. However, within any given species the arabinoside profile (cf. Fig. 2) remained the same regardless of whether or not the walls were isolated from growing or nongrowing oat coleoptiles, soybean hypocotyls, and sycamore-maple suspension cultures.

C. Hydroxyproline Enzymes

Some of the isozymes of horseradish and Japanese radish peroxidase contain hydroxyproline. These isozymes may be specific wall enzymes as they also contain arabinose and galactose (Shannon *et al.*, 1966) and hydroxyproline arabinosides (Liu and Lamport, 1968). This hydroxyproline could be present as a contaminant or as an integral part of the isozyme. If the latter, we must ask whether the hydroxyproline residues are randomly distributed throughout the isozyme or segregated in one area. Does enzymatic attack yield many or few peptides which contain hydroxyproline? According to Morita *et al.* (1968) peptic hydrolysis yielded many peptides, few of which contained hydroxyproline. Therefore, the isozyme contains a hydroxyproline-rich segment. Possibly, this segment is simply a subunit of wall protein, but possibly as an artifactual contaminant. The bona fide presence of such a subunit would allow attachment of the isozyme to the cell wall network. On this basis other wall enzymes should also contain hydroxyproline. This hypothesis is under test. Dr. James Ross in my laboratory recently investigated a wall-bound β-glucosidase some of which can be eluted by 0.5 M NaCl from the cell walls of suspension-cultured tomato cells. The hydroxyproline content of this enzyme increases as the enzyme is purified via salt elution, ammonium sulfate precipitation, and gel filtration on Sephadex G-200 (Table 3). The amounts of enzyme (approximately 60%) which can be eluted from cell walls of young tomato cells (21-day-old cultures) may represent newly formed enzyme which is, at first, ionically bound but which later becomes covalently bound to the wall, as in older cultures only about 10% of the enzyme can be eluted by salt. The significance of *in muro* reactions which these cell surface enzymes catalyze must at present be left pretty much to the imagination of the reader, although some possibilities, such as control of glucose availability by cell wall invertase, have already been investigated (Glasziou, 1969).

IV. ARIADNE'S THREAD

Greek mythology tells of the minotaur, half beast, half human, whose insatiable desire for virginal sacrifice was curbed by Ariadne's strategem. And so after Theseus slew the minotaur in its labyrinthine lair, Ariadne and Theseus retraced their steps guided by Ariadne's silken thread.

Phylogeny is another labyrinth which may, as Huttner (1961) suggested, also have its Ariadne's thread, leading us to common origins. Huttner followed the vitamin B_{12} trail, but that does not lead us through the plant kingdom. On the other hand, if structural hydroxyproline-rich proteins are phylogenetically related, they should lead us through the evolutionary labyrinth toward common origins.

At the level of vertebrates and higher plants the two proteins are grossly dissimilar; collagen consists of a repeating tripeptide: Gly Pro X and a minute amount of carbohydrate (glucosylgalactosylhydroxylysine). In higher plants the hydroxyproline-rich polypeptide component is the minority portion of a glycoprotein where the closest we get to a repeating unit is the common pentapeptide Ser Hyp_4. On tracing our way to the more lowly organisms, additional carbohydrate appears to be attached to collagen. For example, collagen from the sea anemone *Metridium* appears to contain fucose, xylose, mannose, galactosamine, and glucosamine (Katzman and Jeanloz, 1970a) possibly linked via fucosyl-*O*-threonine (Katzman and Jeanloz, 1970b). In the plant kingdom, attempts to follow the thread have been almost restricted to an examination of the hydroxyproline arabinoside profile released by partial alkaline hydrolysis of the cell wall fraction. From tomato down to *Chlorella* this profile is remarkably consistent in hydroxyproline-*O*-arabinosides although the trend seems to be toward shorter arabinose side chains (Table 1). In the more primitive motile green algae, namely *Chlamydomonas* and *Volvox*, there is a striking difference in the hydroxyproline substituents. Here the oligosaccharides attached to hydroxyproline are more varied and larger, as indicated by the hydroxyproline glycoside profile (Fig. 7). Some of these glycosides contain galactose, glucose, and arabinose. We have also tentatively identified hydroxyproline-*O*-galactose in *Chlamydomonas* (D. H. Miller and D. T. A. Lamport, unpublished data) and perhaps in *Volvox*.

Possibly correlated with this dramatic switch to hetero-oligosaccharide substituents is the type of extracellular matrix secreted by these algae. This matrix is not cell wall like; first it tends to disintegrate during isolation, second, in *Volvox*, proteases solubilize the matrix (this ease of solubilization is reminiscent of bacterial peptidoglycan) and third, the formation of new matrix at cell division does not involve fusion of "droplets" along a cell plate growing centrifugally, but is produced after cells

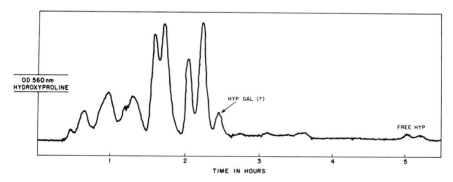

Fɪɢ. 7. Separation of hydroxyproline glycosides from *Volvox*. A Ba(OH)₂ hydrolyzate (0.22 *M*, 6 hours, 100°C) of whole freeze-dried *Volvox carteri* was separated on Chromobeads B as described in the legend to Fig. 1. Note the appearance of peaks additional to those of Fig. 1. These new peaks are hydroxyproline hetero-aligosaccharides.

have divided along an ingrowing cleavage furrow (M. Miller, private communication).

In collaboration with Dr. G. Kochert of the University of Georgia, I recently examined the composition of the protease-solubilized extracellular matrix of *Volvox* (Table 3). Hydroxyproline accounts for 21% of the residues, a value typical for wall proteins. But glycine is exceptionally high (17.6% of total residues), higher in fact than any other wall protein yet analyzed. This might indicate convergence toward a primitive collagen, although there is no *a priori* reason for assuming a high *glycine* content in *primitive* collagen (precollagen?). The presence of hydroxyproline is probably a more primitive feature in view of the fact that its biosynthesis in both plants and animals involves the same cofactors and, most important, involves the fixation of atmospheric oxygen (Prockop *et al.*, 1963; Lamport, 1963). Presumably, the earliest organisms to appear which could utilize molecular oxygen would be those with the ability to generate it. On this basis, hydroxyproline-rich proteins originated among a group of primitive green algae, from which the metazoans ultimately arose. Primitive algae, such as *Cricosphaera* (Aaronson, 1970), which contain wall-bound hydroxyproline, are members of the Chrysophyta. Aaronson (1970) recently suggested that this group contains the ancestral protein from which collagen may have evolved. The Chrysophyta also contain the plant-animals, such as *Ochromonas*, described by Huttner (1961) in *his* search for Ariadne's thread. Whether or not we have followed the thread is a moot point. But somehow we have arrived at the center of the labyrinth and we found *Ochromonas*, the microminotaur.

ACKNOWLEDGMENTS

I am grateful to my colleagues Dr. A. Karr and Dr. J. D. Ross for their unpublished data, to Sandra Roerig and Laura Katona for their exceptionally skilled help, and also to Fred Schroeder and Ed Leinbach for help with the sycamore-maple tryptides. I thank Dr. Clive Bradbeer of the University of Virginia for supplying cultures of *Ochromonas*.

This work was supported by AEC contract AT(11-1)-1338 and NSF grant GB 24361.

REFERENCES

AARONSON, S. (1970). Molecular evidence for evolution in the algae: A possible affinity between plant cell walls and animal skeletons. *Ann N.Y. Acad Sci.* **175**, 531–540.

BOUNDY, J. A., TURNER, J. E., and DIMLER, R. J. (1965). Hydroxyproline-containing mucopolysaccharide from corn pericarp. *Fed. Proc., Fed. Amer. Soc. Exp. Biol.* **24**, 607 (Abstr. No. 2637).

BOUNDY, J. A., WALL, J. S., TURNER, J. F., WOYCHIK, J. H., and DIMLER, R. J. (1967). A mucopolysaccharide containing hydroxyproline from corn pericarp. *J. Biol. Chem.* **242**, 2410–2415.

CHRISPEELS, M. J. (1969). Plant physiol. Synthesis and secretion of hydroxyproline-containing macromolecules in carrots. I. *Kinet. Anal.* **44**, 1187–1193.

CLELAND, R. (1968). Distribution and metabolism of protein-bound hydroxyproline in an elongating tissue, the Avena coleoptile. *Plant Physiol.* **43**, 865–870.

GLASZIOU, K. T. (1969). Control of enzyme formation and inactivation in plants. *Annu. Rev. Plant Physiol.* **20**, 63–88.

GOTELLI, I. B., and CLELAND, R. (1968). Differences in the occurrence and distribution of hydroxyproline-proteins among the algae. *Amer. J. Bot.* **55**, 907–914.

HUTTNER, S. H. (1961). Plant animals as experimental tools for growth studies. *Bull. Torrey Bot. Club.* **88**, 339–349.

JENNINGS, A. C., and WATT, W. B. (1967). Fractionation of plant material. *J. Sci. Food Agr.* **18**, 527–535.

KARR, A. L. (1972). Isolation of an enzyme system which will catalyze glycosylation of "extensin." *Plant Physiol.* **50**, 275–282.

KATZMAN, R. L., and JEANLOZ, R. W. (1970a). Invertebrate connective tissue. *Biochem. Biophys. Res. Commun.* **40**, 628–635.

KATZMAN, R. L., and JEANLOZ, R. W. (1970b). Evidence for a covalent linkage between heteropolysaccharide and collagen from sea anemone. *Fed. Proc., Fed. Amer. Soc. Exp. Biol.* **29**, 599 (Abstract).

LAMPORT, D. T. A. (1963). Oxygen fixation in hydroxyproline of plant cell wall protein. *J. Biochem.* **238**, 1438–1440.

LAMPORT, D. T. A. (1965). Protein component of primary cell walls. *Advan. Bot. Res.* **2**, 151–218.

LAMPORT, D. T. A. (1969). The isolation and partial characterization of hydroxyproline-rich glycopeptides obtained by enzymic degradation of primary cell walls. *Biochemistry* **8**, 1155–1163.

LAMPORT, D. T. A. (1970). Cell wall metabolism. *Annu. Rev. Plant Physiol.* **21**, 235–270.

LAMPORT, D. T. A., and MILLER, D. H. (1971). Hydroxyproline arabinosides in the plant kingdom. *Plant Physiol.* **48**, 454–456.

LIU, E., and LAMPORT, D. T. A. (1968). An hydroxyproline-O-glycosidic linkage in an isolated horseradish peroxidase isozyme. *Plant Physiol.* **43**, S-16.

McBRIDE, O. W., and HARRINGTON, W. F. (1967). Helix-coil transition in collagen. Evidence for a single-stranded triple helix. *Biochemistry* **6**, 1499–1514.

MAIGRET, B., PERAHIA, D., and PULLMAN, B. (1970). Molecular orbital calculations on the conformation of polypeptides and proteins. IV. Conformation of the prolyl and hydroxyprolyl residues. *J. Theor. Biol.* **29**, 275–291.

MORITA, Y., SHIMIZU, K., and KADA, N. (1968). Phytoperoxidase. XX. Peptic hydrolysis of performic acid-oxidized peroxidase a of Japanese radish. *Agr. Biol. Chem.* **32**, 671–677.

OKABAYASHI, H., ISEMURA, T., and SAKAKIBARA, S. (1968). Steric structure of L-proline oligopeptides. II. Far ultraviolet absorption spectra and optical rotations of L-proline oligopeptides. *Biopolymers* **6**, 323–330.

PROCKOP, D., KAPLAN, A., and UNDENFRIEND, S. (1963). Oxygen-18 studies on the conversion of proline to collagen hydroxyproline. *Arch. Biochem. Biophys.* **101**, 499–503.

PUSZTAI, A., and WATT, W. B. (1969). Fractionation and characterization of glycoproteins containing hydroxyproline from the leaves of Vicia faba. *Eur. J. Biochem.* **10**, 523–532.

SHANNON, L. M., KAY, E., and LEW, J. Y. (1966). Peroxidase isozymes from horseradish roots. *J. Biol. Chem.* **241**, 2166–2172.

STEWARD, F. C., ISRAEL, H. W., and SALPETER, M. M. (1967). The labeling of carrot cells with H^3-proline: Is there a cell-wall protein? *Proc. Nat. Acad. Sci. U.S.* **58**, 541–544.

THOMPSON, E. W., and PRESTON, R. D. (1967). Proteins in the cell walls of some green algae. *Nature (London)* **213**, 684–685.

VILLEMEZ, C. L. (1970). Characterization of intermediates in plant cell wall biosynthesis. *Biochem. Biophys. Res. Commun.* **40**, 636–641.

Synthesis and Secretion of Proteins in Plant Cells: the Hydroxyproline-Rich Glycoprotein of the Cell Wall

Maarten J. Chrispeels and David E. Sadava

Department of Biology, John Muir College, University of California at San Diego, La Jolla, California, and Joint Science Department, The Claremont Colleges, Claremont, California

I. INTRODUCTION

Cells of higher plants synthesize a number of extracellular proteins, proteins that need to be transported across the cell membrane (secreted) to arrive at the site of their function. These extracellular proteins include digestive enzymes involved in the hydrolysis of food storage products in cereal grains, structural cell wall proteins, and cell wall enzymes. Research in the area of extracellular proteins has concerned itself primarily with the hormonal regulation of hydrolases synthesized by barley aleurone cells (Varner, 1964; Chrispeels and Varner, 1967a,b; Jacobsen and Varner, 1967; Jones, 1971) and with an analysis of the structure of the hydroxyproline-rich cell wall protein (Lamport, 1965, 1969, this volume). Little research has been done on the mechanisms involved in the biosynthesis and secretion of these extracellular proteins. This gap in our knowl-

edge contrasts with the situation in animal cells, where extracellular proteins have been intensively studied and have been shown to possess a
number of common properties with respect to their structure, synthesis,
and secretion.

Most of the secreted proteins of animal tissues are actually glycoproteins, that is, they have covalently attached sugar residues. This observation led to the hypothesis that glycosylation and secretion are intimately
related (Eylar, 1965): for example, if collagen is not glycosylated within
the cell, it is not secreted (Juva and Prockop, 1966a,b). Moreover, the
Golgi apparatus, which contains some of the glycosylases which attach
sugar residues to the protein backbone (Morré et al., 1969; Schachter et
al., 1970) Wagner and Cynkin, 1971), is also involved in protein secretion
in the exocrine cells of the pancreas (Jamieson and Palade, 1967a,b) and
the goblet cells of the rat colon (Neutra and Leblond, 1966a,b). Another
common property of secreted proteins in animal cells is that they are
synthesized preferentially on membrane-bound polysomes (Redman,
1969), and this often leads to their immediate sequestration within the
cisternae of the endoplasmic reticulum (Redman et al., 1966; Redman,
1967).

With the finding that hydroxyproline is the most abundant amino acid
in a cell wall hydrolyzate (Lamport and Northcote, 1960; Dougall and
Shimbayashi, 1960) and an important consituent of the structural cell
wall glycoprotein (Lamport, 1967), a relatively simple method became
apparent for observing the synthesis and secretion of this component,
namely to label it with radioactive proline and follow the pathway of
protein-bound radioactive hydroxyproline into the wall. This method has
been used successfully to study collagen synthesis and secretion from animal tissues (Rosenbloom and Prockop, 1968). For a plant system, we
used thin disks of phloem parenchyma from the storage roots of the carrot. When such disks are incubated in water (either aseptically or with
chloramphenicol as a bactericide), they undergo a number of physiological biochemical, and cytological changes collectively termed "aging."
This aging process leads to a general increase in respiration (Thimann
et al., 1954), ion uptake (Laties, 1959), and protein synthesis (Ellis and
MacDonald, 1967) and is accompanied by an increase in cytoplasmic
organelles (Fowke and Setterfield, 1968). These changes are dependent
on the synthesis of new messenger RNA (Leaver and Key, 1967). Most
important for our purposes, there is a dramatic increase in the rate of
synthesis of cell wall-bound hydroxyproline. Thus the carrot disk system
provides a homogeneous population of cells which are actively synthesizing and secreting cell wall proteins.

II. BIOSYNTHESIS OF PEPTIDYLHYDROXYPROLINE

A. *Kinetic Experiments*

The first question to be asked about the synthesis of cell wall glycoproteins is at the level of protein synthesis: How does hydroxyproline arise? Early work in Steward's laboratory (Pollard and Steward, 1959) showed that, while carrot cells will take up exogenous radioactive hydroxyproline, they will not incorporate it into proteins; however, labeled proline is not only incorporated into proteins, but also gives rise to labeled hydroxyproline. Furthermore, when tissues are incubated with radioactive proline the biosynthesis of radioactive peptidylhydroxyproline continues even after the labeled amino acid has been removed (Pollard and Steward, 1959) or the radioactivity has been chased with excess nonradioactive proline (Olson, 1964). These experiments suggest that hydroxyproline can be synthesized by plant cells from proline, and that synthesis occurs after the proline is in bound form.

Hydroxyproline synthesis from bound proline could occur at any one of a number of stages of protein synthesis, for example, prolyl-AMP, prolyl-tRNA, peptidylproline in nascent chains, and peptidylproline in completed protein. Hydroxyproline-tRNA has allegedly been isolated from plant cells (Lamport, 1965), but evidence that it is involved in protein synthesis has never been presented. Work with animal cells strongly suggests that hydroxyproline-tRNA does indeed exist but is not functional in protein synthesis (Urivetzky *et al.*, 1965, 1966). This would point to peptidylproline as the precursor to peptidylhydroxyproline, as originally suggested by Steward (Pollard and Steward, 1959).

To show that this is the case *in vivo*, we used the following strategy: the chelating agent α,α'-dipyridyl inhibits the formation of bound hydroxyproline from bound proline but does not have any effect on overall protein synthesis. This inhibition can be reversed by the addition of ferrous ions to the medium (Chrispeels, 1970a). When carrot disks are incubated in [^{14}C]proline in the presence of dipyridyl, the appearance of [^{14}C]hydroxyproline is nearly completely inhibited; however, when ferrous ions are added, there is an initial abrupt increase in the rate of hydroxyproline synthesis (Fig. 1). After several minutes the rate decreases to that observed in the control (no dipyridyl). If just before adding ferrous ions we stop further radioactive proline incorporation by the addition of excess [^{14}C]proline, the initial burst of peptidyl-[^{14}C]hydroxyproline still occurs, but not the subsequent slower accumulation.

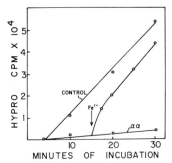

Fig. 1. Reversal of α,α'-dipyridyl-caused inhibition of hydroxyproline (*hypro*) formation by ferrous ions. Aged carrot disks were preincubated with or without dipyridyl (0.15 mM) for 10 minutes and then labeled with [^{14}C]proline. To indicated samples, ferrous ammonium sulfate (0.5 mM) was added. Details and analysis in Chrispeels (1970a). Reprinted with permission.

These data suggest that cells in which hydroxyproline synthesis is inhibited by dipyridyl contain a supply of some proline-rich precursor which can be hydroxylated when ferrous ions are added to remove the inhibition.

That this precursor is peptidylproline can be shown by the experiment reported in Fig. 2. When the protein synthesis inhibitor cycloheximide is added just prior to ferrous ions in an experiment similar to the one just described, protein synthesis stops, but the burst of hydroxyproline synthesis still takes place. The burst can occur 15 minutes after cycloheximide addition. Thus, even when transfer of proline from prolyl-tRNA has been blocked (Felicetti *et al.*, 1966), bound hydroxyproline is synthesized. This clearly indicates that the bound precursor is peptidylproline.

Additional evidence obtained with the same system indicates that hydroxyproline synthesis occurs in the cytoplasm, and not in the cell wall. Addition of Fe^{2+} causes the formation of a constant amount of peptidyl-[^{14}C]hydroxyproline regardless of the length of time the tissue spent in [^{14}C]proline and dipyridyl. Nearly all of this hydroxyproline is in the cytoplasm, and very little of it is associated with the cell wall. If the tissue is labeled in the presence of dipyridyl and then chased in nonradioactive proline for 30 minutes, a sufficient time for secretion (Chrispeels, 1969), the addition of Fe^{2+} causes very little peptidylhydroxyproline synthesis in the wall-bound material.

B. *Peptidylproline Hydroxylase*

Having established that peptidylproline gives rise to peptidylhydroxyproline, we now ask: How does this conversion take place? From *in vivo*

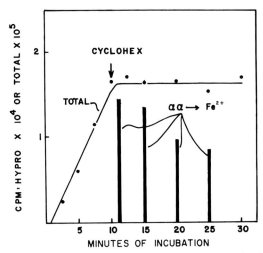

FIG. 2. Effect of cycloheximide (*cyclohex*) on the reversal of hydroxyproline (*hypro*) synthesis inhibition. Aged carrot disks were incubated with dipyridyl and [^{14}C]proline. After 10 minutes, cycloheximide (10 µg/ml) was added. Then, at the times shown, total protein synthesis (curve) or the extent of Fe^{2+}-induced hydroxyproline synthesis (for 5 minutes) were measured (vertical bars). Details in Chrispeels (1970a). Reprinted with permission.

experiments, certain characteristics of an enzyme possibly mediating this conversion are evident. The oxygen atom of the hydroxyl group comes from molecular O$_2$ (Lamport, 1963; Stout and Fritz, 1966). The hydrogen atom on the 4-trans position of proline is displaced by the incoming hydroxyl group (Lamport, 1964). The reaction involves metal ions, probably iron, since dipyridyl inhibits it (see above). These characteristics are to be expected of a hydroxylase (Mason, 1965) and are similar to those exhibited by protocollagen proline hydroxylase in animal tissues (Rosenbloom and Prockop, 1968).

A peptidylproline hydroxylase, fulfilling the above criteria, has recently been characterized from extracts of carrot disks and other plant tissues (Sadava and Chrispeels, 1971). To assay this activity, radioactive proline-rich substrate is prepared by incubating carrot disks in 3,4-*trans*-[^{3}H]proline in the presence of dipyridyl. A specific protein fraction of relative purity (MS, see below) and which normally is extensively hydroxylated is used as substrate. When this substrate is incubated with tissue extracts with necessary cofactors, peptidylproline is converted to peptidylhydroxyproline *in vitro*. The [^{3}H]hydroxyproline formed can be assayed either by a specific assay for the radioactive imino acid (Juva and Prockop, 1966) or by rapidly assaying the tritiated water formed when the 4-*trans*-^{3}H atom of proline is displaced by the incoming hydroxyl group (Hutton *et al.*, 1966).

The characteristics of the plant enzyme are very similar to those of the one in animal cells. It requires molecular oxygen, ferrous ions, ascorbate and α-ketoglutarate for full activity; it does not react with free proline and is not inhibited by free proline or free hydroxyproline. Interestingly, the plant enzyme catalyzes the hydroxylation of prolyl residues of protocollagen (nonhydroxylated collagen) isolated from animal cells; this may point up the lack of specificity of the *in vitro* system, or, possibly, that the plant enzyme does not recognize a proline in a defined primary sequence. This would be anticipated, since a wide variety of plant proteins contain hydroxyproline (Pusztai and Watt, 1969). Peroxidase, which has been implicated in hydroxyproline formation (Yip, 1964), does not copurify with the peptidylproline hydroxylase. By the use of conventional enzymes purification methods, a 75-fold increase in the specific activity over that present in the crude extract has been obtained. The enzyme is found almost exclusively in the soluble cytoplasm (supernatant of 40,000 g for 30 minutes), and very little activity is present in the membranous organelle fraction or the cell wall. Thus, it appears that peptidylhydroxyproline arises from the enzymatic hydroxylation of peptidylproline.

C. Intracellular Site of Hydroxyproline Formation

Does proline hydroxylation occur before or after the precursor polypeptide is released from a polyribosomal protein-synthesizing complex? In animal cells, this question remains unresolved because evidence has been presented on both sides [see Prockop (1970) and Udenfriend (1970) for a summary of the arguments].

Studies on cultured carrot cells (Pollard and Steward, 1959) and cultured tobacco cells (Olson, 1964) indicated that when plant cells incorporate radioactive proline into protein, there is a lag time before the appearance of radioactive hydroxyproline in the proteins. Short-term kinetic analyses on carrot disks show this lag time to be about 3–4 minutes (Chrispeels, 1970b). If the carrot disks are allowed to make this unhydroxylated protein for 3–4 minutes and then excess nonradioactive proline is added, the normal level of hydroxylation of proline is reached within 10 minutes (Sadava and Chrispeels, 1971a); presumably this chase period is well beyond the time needed to release the completed radioactive proteins from polyribosomes (Winslow and Ingram, 1966). If the release of nascent chains is blocked by cycloheximide, hydroxylation of proline is severely inhibited: In the experiment reported in Table 1, cyclohexi-

TABLE 1
EFFECT OF CYCLOHEXIMIDE AND PUROMYCIN ON
HYDROXYPROLINE FORMATION[a]

| Time of incubation (minutes) | % [14C]Proline residues hydroxylated | |
	Control	Antibiotic
		Cycloheximide
10	10.6	3.2
30	14.2	4.5
60	17.8	4.5
90	20.0	4.6
		Puromycin
20	14.0	15.5

[a] Aged carrot disks were preincubated with cyclohex-imide (100 μg/ml) or puromycin (100 μg/ml) and subsequently labeled with [14C]proline for the times indicated. The extent of [14C]proline hydroxylation in the proteins (cytoplasm + cell wall) was measured as in Chrispeels (1969).

mide inhibited the incorporation of [14C]proline into protein by 95%; the proline which was incorporated under these conditions was not fully hydroxylated, and the biosynthesis of hydroxyproline was inhibited by 75% (over and above the inhibition of protein synthesis). The incorporated radioactivity was presumably largely in nascent proteins. In the presence of puromycin, hydroxyproline biosynthesis is not inhibited; this antibiotic inhibits protein synthesis by detaching the nascent proteins from the polyribosomes. Finally, in the reversal experiment noted above, the burst of hydroxyproline synthesis observed when Fe^{2+} was added to dipyridyl-inhibited disks is still observed after a chase in nonradioactive proline. In sum, the kinetic data are consistent with the idea that peptidylproline becomes hydroxylated after the nascent protein is released into the cytoplasm.

That this is the case is borne out by our analyses of polyribosome-associated proteins presumed to be nascent chains (Sadava and Chrispeels, 1971a) If carrot disks are incubated in radioactive proline ribosomes isolated and fractionated on sucrose density gradients, most of the bound radioactivity is associated with the heavy, polyribosomal region of the gradient. Analysis of this bound radioactivity by two methods—chromatographic separation of proline and hydroxyproline after acid hydrolysis of the protein (Chrispeels, 1969), and specific assay of hydroxyproline

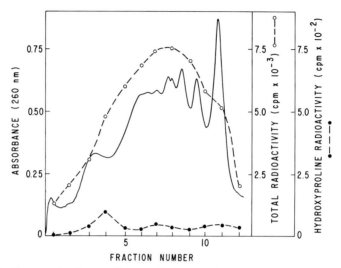

FIG. 3. Sucrose density gradient analysis of polysomes. Aged carrot disks were incubated for 10 minutes with [¹⁴C]proline. Polyribosomes were isolated from the membranous organelle fraction of the cytoplasm and analyzed on a 10–34% sucrose gradient. Absorbance was determined on a continuous recording spectrophotometer and bound radioactivity assayed for hydroxyproline. Details in Sadava and Chrispeels (1971a).

after acid hydrolysis of the protein (Juva and Prockop, 1966a)—shows that little or no proline is hydroxylated (Fig. 3). In the experiments disks were allowed to incorporate [¹⁴C]proline for 10 minutes. Under these conditions about 10% of the proline residues in the soluble cytoplasm and 15% in the membranous organelles are hydroxylated. However, less than 1% of the radioactivity associated with the polyribosomes isolated from the soluble cytoplasm or the membranous organelles could be accounted for as hydroxyproline.

Our contention that this radioactivity bound to the polyribosomal region represents nascent protein is based on the following lines of evidence:

1. Incubation of ribosomal suspensions in ribonuclease shows, upon gradient analysis, that the ultraviolet-absorbing material and associated radioactivity in the polyribosome region disappears and there is a concomitant increase in material in the monoribosome region; this would be expected for ribosomal aggregates stabilized by messenger RNA.

2. Kinetics of incorporation show that polysome-associated radioactivity reaches a plateau in a short (5–7 minutes) time and can be chased from the region by 7 minutes' incubation in nonradioactive proline; this

is to be expected if protein synthesis takes a short time (Winslow and Ingram, 1966).

3. Gel filtration on Sephadex G-100 of polypeptides released from the ribosomes by alkaline hydrolysis (breaks the bond between tRNA and nascent chains) shows a wide size distribution, indicating that not only short peptides but molecules at various stages of completion are present.

Similar experiments performed on cultured tobacco cells give the same result: bound radioactivity associated with polyribosomes contains little or no hydroxyproline.

Thus, while very limited hydroxylation on nascent chains cannot be completely ruled out, the evidence indicates that the precursor to peptidylhydroxyproline is the protein released from a ribosomal complex.

III. GLYCOSYLATION OF PEPTIDYLHYDROXYPROLINE

Our understanding of the structure of hydroxyproline-rich wall protein was significantly advanced with the discovery that the hydroxyproline residues are glycosylated with arabinose through an O-glycosidic linkage involving the hydroxyl group of the hydroproline (Lamport, 1967). When cell walls are subjected to alkaline hydrolysis it is possible to isolate small glycopeptides which consist of a hydroxyproline residue attached to a short chain of arabinose residues. Similar glycopeptides can be isolated from cytoplasmic hydroxyproline-rich proteins (Fig. 4) which are on their way to the cell wall; this suggests that the attachment of arabinose occurs in the cytoplasm, not after the protein arrives in the cell wall. In the experiment shown, the tissue was labeled with either [^{14}C]proline or [^{14}C]arabinose and the proteins from the membranous organelles were hydrolyzed in alkali and the neutralized hydrolyzate was chromatographed on a cation exchange column. Four peaks (Nos. 1 through 4) appeared which contained both arabinose and hydroxyproline; these have been characterized as hydroxyproline-(arabinose)$_n$ where n is 4, 3, 2 and 1, respectively (Lamport, 1969). Peaks 5 and 6 represent epimers of free hydroxyproline and constitute the unglycosylated hydroxyproline.

The presence of hydroxyproline O-arabinosides in the cytoplasm suggests that hydroxylation and glycosylation are sequential processes which precede secretion of the glycoprotein into the cell wall. That this is indeed the case can be shown by following the kinetics of the incorporation of radioactive proline into proteins, the appearance of radioactive bound hydroxyproline, and the appearance of radioactive glycosylated hydroxyproline (Fig. 5). It is clear that proline hydroxylation lags behind protein

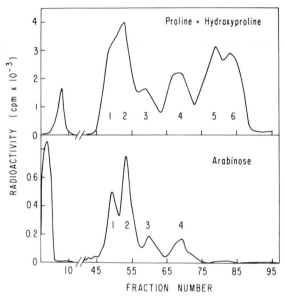

Fig. 4. Hydroxyproline arabinosides isolated from cytoplasmic proteins. Aged carrot disks were incubated with [^{14}C]proline or [^{14}C]arabinose for 30 minutes. Membranous organelles were prepared (Chrispeels, 1969), dialyzed, and hydrolyzed in 0.4 M Ba(OH)$_2$ for 18 hours at 120°C. The hydrolyzate was neutralized, cleared, adjusted to pH 3.0, and adsorbed on a cation exchange column which was eluted with a 0–0.2 N HCl gradient (250 ml each). Radioactivity in each 5-ml fraction was determined on a 1-ml aliquot. [^{14}C]Proline was not eluted from the column under these conditions.

synthesis by 3–4 minutes and that glycosylation follows another 3–4 minutes later. The glycosylated protein is then secreted 12–15 minutes after it was synthesized.

The characterization of the hydroxyproline-(arabinose)$_n$ glycopeptides and of the glycosylases involved in their synthesis are discussed elsewhere in this volume.

IV. SECRETION OF THE GLYCOPROTEIN

Plant cells, unlike animal cells, possess a variety of proteins which contain hydroxyproline (Steward and Chang, 1963; Pusztai and Watt, 1969). Subsequent to the discovery of hydroxyproline in plant cells (Steward *et al.*, 1951) and the pioneering work in Steward's laboratory, it was reported that a large proportion of the protein-bound hydroxypro-

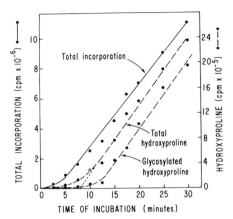

FIG. 5. Kinetics of glycosylation of peptidylhydroxyproline. Aged carrot disks were labeled with [¹⁴C]proline for the times indicated and then homogenized. Aliquots of the homogenate were used to determine total incorporation, radioactivity in peptidylhydroxyproline (after acid hydrolysis of the proteins), radioactivity in peptidylhydroxyproline arabinoside (after alkaline hydrolysis of the proteins). Details in Chrispeels (1970b). Reprinted with permission.

line is associated with a cell wall preparation (Lamport and Northcote, 1960; Dougall and Shimbayashi, 1960). It has since become a matter of some controversy whether protein-bound hydroxyproline is really localized in the cell wall *in situ* (*in muro*) or whether it becomes associated with the wall during homogenization (Lamport, 1965; 1970; Israel *et al.*, 1968; Steward *et al.*, 1970). Our own evidence, obtained primarily with carrot disks (Sadava and Chrispeels, 1969; Chrispeels, 1969), supports the idea that the cell wall contains hydroxyproline-rich proteins. However, it is now well established that hydroxyproline-containing proteins are also present in the cytoplasm (Steward and Chang, 1963; Cleland, 1968; Sawai and Morita, 1970).

Evidence obtained in our own laboratory can be summarized as follows. First, when aged carrot disks are homogenized most of the hydroxyproline is found in the cell wall fraction. As aging proceeds, more and more hydroxyproline accumulates in this fraction, but little change occurs in the cytoplasm (Chrispeels, 1969). Extensive washing of the cell wall fraction with salt, detergent, or water does not remove the hydroxyproline. Second, when carrot disks are allowed to incorporate radioactive proline for 24 hours, then plasmolyzed to separate the cell wall from the cytoplasm and subsequently prepared for autoradiography, approximately one-fourth of the silver grains are found over the walls and the rest over the cytoplasm (Fig. 6). Third, pulse-chase experiments clearly indicate

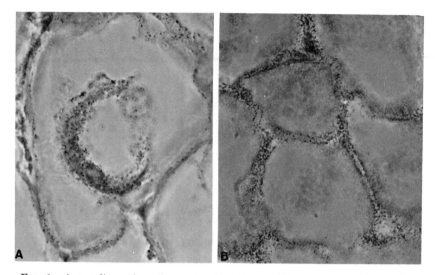

Fig. 6. Autoradiography of carrot cells. Carrot disks were aged for 24 hours and labeled with tritiated proline (randomly labeled, 5 μCi per disk) for 24 hours. The disks were washed, plasmolyzed, fixed, and embedded. Mounted sections were coated with emulsion, exposed for 3 days at 4°C, and developed. (A) Focused on cell structures (× 850). (B) Focused on cell wall (× 430). Reprinted with permission from Sadava and Chrispeels (1969).

that the synthesis and the accumulation of hydroxyproline occur in two distinct cellular compartments, one in the cytoplasm and one in, or associated with, the cell wall.

The timing of secretion to the cell wall can be inferred by first incubating a tissue in radioactive proline, then allowing a chase in nonradioactive proline and observing the amount of bound radioactive hydroxyproline proteins in various cell fractions. When this is done on a crude scale with oat coleoptiles (Cleland, 1968) or cultured tobacco cells (Olson, 1964), there is a decrease in cytoplasmic hydroxyproline and a concomitant increase in the wall fraction during the chase period. More exact short-term kinetics with carrot disks (Fig. 7) indicate that the glycosylated hydroxyproline protein arrives in the cell wall about 5 minutes after glycosylation (13 minutes after proline incorporation). This rapid secretion contrasts with that in animal cells: in pancreatic cells, the time from initial incorporation to secretion of hydrolases is 1 hour (Jamieson and Palade, 1967a); for collagen, it is 30 minutes (Bhatnagar et al., 1967).

In many animal systems, as was noted above, secretion of proteins is mediated by membranous organelles. When the cytoplasm of carrot disks incubated in radioactive proline is fractionated into "membranous

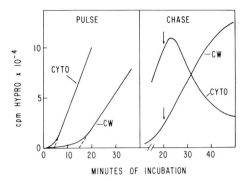

FIG. 7. Appearance and disappearance of protein-bound hydroxyproline (*hypro*) in the cytoplasm (*cyto*) and the cell wall (*CW*) during pulse-chase experiments with [^{14}C]proline. Aged carrot disks were labeled with [^{14}C]proline and harvested at 5-minute intervals. Appearance of peptidyl-[^{14}C]hydroxyproline in the cytoplasm and the cell wall is shown in the left-hand panel. In some samples the radioactivity was chased with 0.01 M [^{12}C]proline and samples were again harvested at 5-minute intervals and peptidyl-[^{14}C]hydroxyproline was measured in the cytoplasm and the cell wall (right-hand panel).

organelle" and "supernatant" fractions, all of the peptidyl-[^{14}C]hydroxy-proline which is to be secreted is associated with the membranous organelles (Chrispeels, 1969). It is not clear whether the proteins are contained in specific organelles or merely bound to their outside surfaces. In cultured sycamore cells, the turnover of bound hydroxyproline as measured by pulse-chase kinetics is most rapid in a fraction containing both dictyosomes and smooth endoplasmic reticulum, and this turnover can account for the arrival of this component in the cell wall (Dashek, 1970). An electron micrograph of this organelle fraction is shown in Fig. 8.

Recent evidence on the localization of glycosyl transferase enzymes involved in glycoprotein synthesis in animal cells (Morré *et al.*, 1969; Schachter *et al.*, 1970; Wagner, and Cynkin, 1971) and carbohydrate synthesis in plant cells (Ray *et al.*, 1969) suggest that the Golgi apparatus and the smooth endoplasmic reticulum may be simultaneously involved in the biosynthesis of glycoproteins and their secretion.

In many plant systems which secrete proteins it is difficult to separate protein synthesis from protein secretion (Chrispeels and Varner, 1967a). Often there is no substantial intracellular accumulation of proteins. Pulse-chase experiments with [^{14}C]proline make it possible to study the requirements for the secretion of hydroxyproline-rich proteins separate from the requirements for synthesis. It is possible, for example, to add

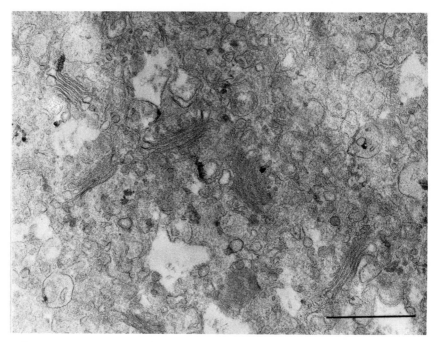

Fig. 8. Membranous organelles (isolated by discontinuous sucrose gradient centrifugation) involved in the transport of hydroxyproline-rich proteins from the cytoplasm to the cell wall in cultured sycamore-maple cells. ×35,000. Reprinted with permission of Dashek (1970).

inhibitors of specific metabolic reactions immediately after the start of the chase and then study their effects on the disappearance of labeled proteins from the cytoplasm or their appearance in the wall. Such experiments show that the secretion of hydroxyproline-rich proteins is an active process which is dependent on metabolic energy, probably in the form of ATP (Doerschug and Chrispeels, 1970). Uncouplers of oxidative phosphorylation (m-chlorocarbonyl cyanide) phenylhydrazone, and p-trifluoromethoxy (carbonyl cyanide) phenylhydrazone, completely inhibit the secretory process whereas, inhibitors of electron transport, such as cyanide and azide, cause a substantial inhibition (Doerschug and Chrispeels, 1970). Similar observations have been made with animal cells (Jamieson and Palade, 1968; Babad et al., 1967; Bauduin et al., 1969) and with other plant cells (Chrispeels and Varner, 1967a; Jones, 1971).

Secretion of these cell wall proteins is not dependent on concomitant protein synthesis, and the addition of cycloheximide at the time of the chase does not inhibit the transport of the proteins to the wall (Doerschug and Chrispeels, 1970).

V. CHARACTERIZATION OF A CELL WALL GLYCOPROTEIN AND ITS CYTOPLASMIC PRECURSOR

Because most of the hydroxyproline-rich glycoproteins are covalently attached to the cell wall matrix, isolation of an intact wall glycoprotein is not possible. Generally, analysis is made of walls digested by chemical and enzymatic methods (Lamport, 1969), and, while this gives considerable information on the molecular architecture of the wall, it adds little to our understanding of the assembly and secretion of a given wall component.

One approach toward the isolation of an intact wall glycoprotein is to isolate its cytoplasmic precursor on the basis of pulse-chase kinetics and then look for the same molecule while it is associated with, but not yet covalently linked to, the wall. The hydroxyproline-containing glycoproteins which are transported from the cytoplasm to the cell wall are transiently associated with the membranous organelles of carrot disks. One component of this fraction, soluble in cold 5% trichloroacetic acid, accounts for about 30% of all the peptidylhydroxyproline on its way to the wall (Chrispeels, 1969). Not all the hydroxyproline macromolecules are covalently attached to the wall; some are extractable with 0.2 M salt, and this component also contains trichloroacetic acid-soluble material. Since acid solubility affords considerable purification, further analysis has been made of the cytoplasmic (MS) and cell wall (CWS) material (M. Brysk and M. J. Chrispeels, 1971, unpublished observations).

Kinetic data are consistent with the MS-CWS representing a cell wall glycoprotein. The MS becomes maximally labeled with exogenous proline after 30 minutes and has a half-life upon chase of 11–12 minutes, which would be expected given the rapid rate of secretion mentioned above. The CWS only becomes labeled after 20 minutes, and continues to be labeled during chase, also expected if it is a cell wall protein. Moreover, as more radioactive macromolecules are built into the wall covalently, during chase the proportion of the total wall radioactivity in the CWS declines from 43% initially to 18% after 3 hours. Finally, about 80% of the radioactivity in the MS or CWS is in hydroxyproline, the remainder in proline.

The MS and CWS cochromatograph on carboxymethyl Sephadex and exhibit identical physical and chemical properties. Some of these properties are shown in Table 2. It is difficult at present to reconcile the sedi-

TABLE 2
PHYSICAL AND CHEMICAL PROPERTIES OF INTACT WALL GLYCOPROTEIN

Property	Method for determination	Result
Molecular weight	Sephadex G-200	Approximately 200,000
Sedimentation coefficient	Sucrose gradient	Approximately 4.5 S
Isoelectric point	Electrophoresis	pH 9.5
Density	Isopycnic CsCl gradient	1.45 gm/cc
Sugars	Thin-layer chromatography	Arabinose
		Galactose, rhamnose (traces)
Amino acids	Amino acid analyzer	Mole % Hyp 10.3
		Pro 2.8
		Ser 18.8
		Gly 12.0
		Lys 12.2

mentation coefficient with the molecular weight estimation: globular proteins with a sedimentation coefficient of 4.5 S have molecular weights of about 50,000. Perhaps the glycoprotein is highly asymmetric. The density determination indicates that the glycoprotein is 60% protein (density 1.30) and 40% carbohydrate (density 1.65). The results of the sugar analyses can be qualitatively confirmed by the observation that the MS-CWS becomes readily labeled with radioactive arabinose and slightly labeled with galactose. Amino acid analysis with a ratio of 4:1 for hydroxyproline:proline is identical with the ratio of the radioactive amino acids in the crude MS; the presence of high amounts of other hydroxyamino acids is similar to the situation found in cell wall digests (Lamport, 1965).

In summary, the isolated MS-CWS shows all of the kinetic and physicochemical properties to be expected of an intact cell wall glycoprotein. A remaining unanswered question is how this glycoprotein is built into the wall matrix.

VI. REGULATION OF CELL WALL PROTEIN SYNTHESIS

What is the role of the hydroxyproline-rich protein in the cell wall and what are the consequences of the synthesis and secretion of these proteins to the physiology of the cell?

The hydroxyproline-rich glycoprotein is obviously a structural part of the wall. It has been named "extensin," and it was suggested that it may play an important role in regulating the extension of the wall which accompanies cell elongation. The hydroxyproline content of the wall increases as cells elongate (Cleland, 1968; Cleland and Karlsnes, 1967). However, in pea stems most of the wall hydroxyproline is laid down after the cessation of elongation (Cleland and Karlsnes, 1967; Ridge and Osborne, 1970). Inhibition of hydroxyproline formation by chelators leads to a temporary increase in the growth rate and an increase in wall extensibility (Barnett, 1970). These results are all consistent with the idea that in some tissues "extensin" may play a role in cross-linking and stiffening the wall at the cessation of cell expansion.

We can approach this problem by studying the synthesis of wall hydroxyproline under various conditions of growth. Excision of carrot disks followed by incubation leads to a 10-fold increase in the hydroxyproline content of the cell wall (Chrispeels, 1969). This increase is not a peculiarity of carrot disks or of the incubation conditions. When small pieces of carrot phloem parenchyma are incubated aseptically in White's medium (White, 1943), supplemented with 10% coconut milk, they greatly increase in size, fresh weight, and dry weight over a period of several weeks (Steward et al., 1964). Under these conditions a similar rise in the hydroxyproline content of the wall takes place. The same phenomenon is observed when small disks of tobacco pith are incubated on nutrient media which either do not or do induce cell proliferation and callus formation. It appears that excision of the tissue followed by incubation is all that is necessary to enhance the synthesis of the hydroxyproline-rich protein with respect to the other cell wall components. This is in agreement with the observations that cells growing in an undifferentiated non-self-limiting manner contain more hydroxyproline (Steward et al., 1958; Lamport, 1965). If, however, such cells are induced to resume normal differentiated growth by proper hormonal balance, the wall hydroxyproline content returns to the lower level of the parent tissue. This is shown in Table 3: undifferentiated callus derived from tobacco stems (seedlings) had a greatly increased wall hydroxyproline level, but when differentiation was induced the level returned to normal.

How does the plant cell achieve this regulation of wall protein synthesis? The change in wall hydroxyproline upon aging in carrot disks is accompanied by an increase in the extent of proline hydroxylation as measured by isotopic experiments. In unaged disks less than 1% of their proline residues is hydroxylated; after 20 hours of aging, the extent is 25%. A similar situation holds for pea stems, which accumulate wall hydroxyproline after cell elongation has ceased (Cleland and Karlsnes, 1967); in this case the extent of hydroxylation rises from 18% in elongating tissue to 33% in the mature stem. These results suggest a change in the capacity to synthesize peptidylhydroxyproline. There is under

TABLE 3

HYDROXYPROLINE CONTENTS OF CELL WALLS FROM
DIFFERENT TOBACCO TISSUES[a]

Nature of tissue	Hydroxyproline/cell wal (μg/mg)
Seedling stem	2.7
Callus	14.0
Malformed shoots (10 mg/liter cytokinin)	4.7
Small leaves (0.1 mg/liter cytokinin)	2.0
Large leaves (1.0 mg/liter cytokinin)	2.0
Pith from mature plants	1.8

[a] Small stem sections of 2-week-old tobacco seedlings (var. Wisconsin 38) were transplanted aseptically to a nutrient medium containing salt, vitamins, sucrose indoleacetic acid (2 mg/ml), and cytokinin, and 4 weeks later the different tissues were harvested. The cell walls were isolated, purified, dried and weighed, hydrolyzed and their hydroxyproline contents determined colorimetrically.

these conditions no change in the specific activity of the peptidylproline hydroxylase (D. Sadava, unpublished). Regulation apparently does not occur at this level. By measuring the extent of secretion of hydroxyproline proteins, it can be shown that much more of the synthesized peptidylhydroxyproline is secreted as aging proceeds; this is true in carrot disks and in pea stems which have ceased elongation. There appears to be an increase in the rate at which cell wall proteins are synthesized and secreted. This increase is blocked by actinomycin D, suggesting that the regulation may be achieved by making more messenger RNA molecules for hydroxyproline-rich cell wall protein available for translation.

The role of these glycoproteins in the wall still remains very much a mystery a decade after their discovery. But this problem will no doubt be investigated in a number of laboratories. We consider cell wall hydroxyproline to be the pectin methylesterase of the 1970's. Whether or not it has a regulatory role in cell growth remains to be discovered, claims to the contrary notwithstanding (Lamport, this volume)

ACKNOWLEDGMENTS

It is a pleasure to acknowledge the help of my collaborators Miriam Brysk, Marcia Doerschug, and David Sadava. Each of them contributed significantly to the work reported in this paper. Dr. C. Y. Lin gave us valuable assistance with the problem of isolating polysomes. My research has benefited from a number of interesting discussions about protein synthesis and protein secretion with Tom Humphreys and Joe Varner. The work described in this paper has been supported primarily by the United States Atomic Energy Commission under Contract AT(04-3)-34; Project Agreement 159.

REFERENCES

BADAD, H., BEN-ZVI, R., BDOLAH, A., and SCHRAMM, M. (1967). The mechanism of enzyme secretion by the cell. *Eur. J. Biochem.* **1**, 96–101.

BARNETT, N. M. (1970). Dipyridyl-induced cell elongation and inhibition of cell wall hydroxyproline biosynthesis. *Plant Physiol.* **45**, 188–191.

BAUDIN, H., COLIN, M., and DUMONT, J. E. (1969). Energy sources for protein synthesis and enzymatic secretion in rat pancreas *in vitro*. *Biochim. Biophys. Acta* **174**, 722–733.

BHATNAGAR, R. S., KIVIRIKKO, K. I., ROSENBLOOM, J., and PROCKOP, D. J. (1967). Transfer of puromycin-containing polypeptides through the plasma membrane of cartilage cells synthesizing collagen. *Proc. Nat. Acad. Sci. U.S.* **58**, 248–255.

CHRISPEELS, M. J. (1969). Synthesis and secretion of hydroxyproline containing macromolecules in carrots. I. Kinetic analysis. *Plant Physiol.* **44**, 1187–1193.

CHRISPEELS, M. J. (1970a). Synthesis and secretion of hydroxyproline containing macromolecules in carrots. II. *In vivo* hydroxylation of peptidyl proline to peptidyl hydroxyproline. *Plant Physiol.* **45**, 223–227.

CHRISPEELS, M. J. (1970b). Biosynthesis of cell wall protein: Sequential hydroxylation of proline, glycosylation of hydroxyproline and secretion of the glycoprotein. *Biochem. Biophys. Res. Commun.* **39**, 732–737.

CHRISPEELS, M. J., and VARNER, J. E. (1967a). Gibberellic acid-enhanced synthesis and release of amylase and ribonuclease by isolated aleurone layers. *Plant Physiol.* **42**, 398–406.

CHRISPEELS, M. J., and VARNER, J. E. (1967b). Hormonal control of enzyme synthesis: On the mode of action of giberrellic acid and abscisin in aleurone layers of barley. *Plant Physiol.* **42**, 1008–1016.

CLELAND, R. (1968). Distribution and metabolism of protein-bound hydroxyproline in an elongating tissue, the *Avena coleoptile*. *Plant Physiol.* **43**, 865–870.

CLELAND, R., and KARLSNES, A. M. (1967). A possible role of hydroxyproline-containing proteins in the cessation of cell elongation. *Plant Physiol.* **42**, 669–671.

DASHEK, W. V. (1970). Synthesis and transport of hydroxyproline-rich components in suspension cultures of Sycamore-Maple cells. *Plant Physiol.* **46**, 831–838.

DOERSCHUG, M. R., and CHRISPEELS, M. J. (1970). Synthesis and secretion of hydroxyproline containing macromolecules in carrots. III. Metabolic requirements for secretion. *Plant Physiol.* **46**, 363–366.

DOUGALL, D. K., SIMBAYASI, K. (1960). Factors affecting growth of tobacco callus tissue and its incorporation of tyrosine. *Plant Physiol.* **33**, 396–404.

ELLIS, R. J., and MACDONALD, I. R. (1967). Activation of protein synthesis by microsomes from aging beet disks. *Plant Physiol.* **42**, 1297–1302.

EYLAR, E. H. (1965). On the biological role of glycoproteins. *J. Theor. Biol.* **10**, 89–113.

FELICETTI, L., COLOMBO, B., and BAGLIONI, C. (1966). Inhibition of protein synthesis in reticulocytes by antibiotics. II. The site of action of cycloheximide, streptovitacin A and pactamycin. *Biochim. Biophys. Acta* **119**, 120–129.

FOWKE, K., and SETTERFIELD, G. (1968). Cytological responses of Jerusalem artichoke tuber slices during ageing and subsequent auxin treatment. *In* "Biochemistry and Physiology of Plant Growth Substances" (F. Wightman and G. Setterfield, eds.). Runge Press, Ottawa.

HUTTON, J. J. JR., TAPPEL, A. L., and UDENFRIEND, S. (1966). Cofactor and substrate requirement of collagen proline hydroxylase. *Arch. Biochem. Biophys.* **118**, 231–240.

ISRAEL, H. W., SALPETER, M. M., and STEWARD, F. C. (1968). The incorporation of radioactive proline into cultured cells. *J. Cell Biol.* **39**, 698–715.

JACOBSEN, J. V., and VARNER, J. E. (1967). Gibberellic acid-induced synthesis of protease by isolated aleurone layers of barley. *Plant Physiol.* **42**, 1596–1600.

JAMIESON, J. D., and PALADE, G. E. (1967a). Intracellular transport of secretory proteins in the pancreatic exocrine cell. I. *J. Cell Biol.* **34**, 577–596.

JAMIESON, J. D., and PALADE, G. E. (1967b). Intracellular transport of secretory proteins in the pancreatic exocrine cell. II. *J. Cell Biol.* **34**, 597–615.

JAMIESON, J. D., and PALADE, G. E. (1968). Intracellular transport of secretory proteins in the pancreatic exocrine cell. IV. Metabolic requirements. *J. Cell Biol.* **39**, 584–603.

JONES, R. L. (1971). Gibberellic acid enhanced release of β-1,3 glucanase from barley aleurone cells. *Plant Physiol.* **47**, 412–415.

JUVA, K., and PROCKOP, D. J. (1966a). Modified procedure for the assay of H³- or C¹⁴-labeled hydroxyproline. *Anal. Biochem.* **15**, 77–83.

JUVA, K., and PROCKOP, D. J., (1966b). Hydroxylation of proline and the intracellular accumulation of a polypeptide precursor of collagen. *Science* **152**, 92–94.

LAMPORT, D. T. A. (1963). Oxygen fixation into hydroxyproline of plant cell wall protein. *J. Biol. Chem.* **238**, 1438–1440.

LAMPORT, D. T. A. (1964). Hydroxyproline biosynthesis: Loss of hydrogen during the hydroxylation of proline. *Nature (London)* **202**, 293–294.

LAMPORT, D. T. A. (1965). The protein component of primary cell walls. *Advan. Bot. Res.* **2**, 151–218.

LAMPORT, D. T. A. (1967). Hydroxyproline-O-glycosidic linkage of the plant cell wall glycoprotein extensin. *Nature (London)* **216**, 1322–1324.

LAMPORT, D. T. A. (1969). The isolation and partial characterization of hydroxyproline-rich glycopeptides obtained by enzymic degradation of primary cell walls. *Biochemistry* **8**, 1155–1163.

LAMPORT, D. T. A. (1970). Cell wall metabolism. *Annu. Rev. Plant Physiol.* **21**, 235–270.

LAMPORT, D. T. A., and Northcote, D. H. (1960). Hydroxyproline in primary cell walls of higher plants. *Nature (London)* **188**, 665–666.

LATIES, G. G., (1959). Active transport of salt into plant tissue. *Annu. Rev. Plant Physiol.* **10**, 87–112.

LEAVER, C. J., and KEY, J. L. (1967). Polyribosome formation and RNA synthesis during aging of carrot-root tissue. *Proc. Nat. Acad. Sci. U.S.* **57**, 1338–1344.

MASON, H. S. (1965). Oxidases. *Annu. Rev. Biochem.* **34**, 595–634.

MORRÉ, D. J., MERLIN, L. M., and KEENAN, T. W. (1969). Localization of glycosyl transferase activities in a Golgi apparatus-rich fraction isolated from rat liver. *Biochem. Biophys. Res. Commun.* **37**, 813–819.

NEUTRA, M., and LEBLOND, C. P. (1966a). Synthesis of the carbohydrate of mucus in the Golgi complex as shown by electron microscope radioautography of goblet cells from rats injected with glucose-³H. *J. Cell Biol.* **30**, 119–136.

NEUTRA, M., and LEBLOND, C. P. (1966b). Radioautographic comparison of the uptake of galactose-³H and glucose-³H in the Golgi region of various cells secreting glycoproteins of mucopolysaccharides. *J. Cell Biol.* **30**, 137–150.

OLSON, A. C. (1964). Proteins and plant cell walls. Proline to hydroxyproline in tobacco suspension cultures. *Plant Physiol.* **39**, 543–550.

POLLARD, J. K., and STEWARD, F. C. (1959). The use of C¹⁴-proline by growing cells: Its conversion to protein and to hydroxyproline. *J. Exp. Bot.* **10**, 17–32.

PROCKOP, D. J. (1970). Intracellular biosynthesis of collagen and interactions of protocollagen proline hydroxylase with large polypeptides. *In* "Chemistry and Molecular Biology of the Intercellular Matrix" (E. A. Balazs, ed.), Vol. 1, pp. 335–370. Academic Press, New York.

PUSZTAI, A., and WATT, W. B. (1969). Fractionation and characterization of glycoproteins containing hydroxyproline from the leaves of *Vicia faba*. *Biochemistry* **10**, 523–532.

RAY, P. M., SHININGER, T. L., and RAY, M. M. (1969). Isolation of β-glucan synthetase particles from plant cells and identification with Golgi membranes. *Proc. Nat. Acad. Sci. U.S.* **64**, 605–612.

REDMAN, C. M. (1967). Studies on the transfer of incomplete polypeptide chains across rat liver microsomal membranes *in vitro*. *J. Biol. Chem.* **242**, 761–768.

REDMAN, C. M. (1969). Biosynthesis of serum proteins and ferritin by free and attached ribosomes of rat liver. *J. Biol. Chem.* **244**, 4308–4315.

REDMAN, C. M., SIEKEVITZ, P., and PALADE, G. E. (1966). Synthesis and transfer of amaylase in pigeon pancreatic microsomes. *J. Biol. Chem.* **241**, 1150–1158.

RIDGE, I., and OSBORNE, D. J. (1970). Hydroxyproline and peroxidase in cell walls of *Pisum sativum*: Regulation by ethylene. *J. Exp. Bot.* **21**, 843–856.

ROSENBLOOM, J., and PROCKOP, D. J. (1968). Biochemical aspects of collagen biosynthesis. *In* "Repair and Regeneration: Scientific Basis for Medical Practice" (J. E. Dunphy and H. W. Van Winkle, eds.), pp. 117–135. McGraw-Hill, New York.

SADAVA, D., and CHRISPEELS, M. J. (1969). Cell wall protein in plants: Autoradiographic evidence. *Science* **165**, 299–300.

SADAVA, D., and CHRISPEELS, M. J. (1971). Hydroxyproline biosynthesis in plant cells. Peptidyl proline hydroxylase from carrot disks. *Biochim. Biophys. Acta* **227**, 278–287.

SADAVA, D., and CHRISPEELS, M. J. (1971a). *Biochemistry* **10**, 4290–4294.

SAWAI, H., and MORITA, Y. (1970). Studies on γ₁-globulin. *Agr. Biol. Chem.* **34**, 61–67.

SCHACHTER, H., JABBAL, I., HUDGIN, R. L., PINTERIC, L., McGUIRE, E. J., and ROSEMAN, S. (1970). Intracellular localization of liver sugar nucleotide glycoprotein glycosyltransferases in a Golgi-rich fraction. *J. Biol. Chem.* **245**, 1090–1100.

STEWARD, F. C., and CHANG, L. O. (1963). The incorporation of ¹⁴C-proline into the proteins of growing cells. II. A note on evidence from cultured carrot explants by acrylamide gel electrophoresis. *J. Exp. Bot.* **14**, 379–386.

STEWARD, F. C., THOMPSON, J. F., MILLAR, F. K., THOMAS, M. D., and HENDRICKS, R. H. (1951). The amino acids of alfalfa as revealed by paper chromatography with special reference to compounds labelled with S³⁵. *Plant Physiol.* **26**, 123–135.

STEWARD, F. C., THOMPSON, J. F., and POLLARD, J. K. (1958). Contrasts in the nitrogenous composition of rapidly growing and non-growing plant tissues. *J. Exp. Bot.* **9**, 1–10.

152   MAARTEN J. CHRISPEELS AND DAVID E. SADAVA

STEWARD, F. C., MAPES, M. O., KENT, A. E., and HOLSTEN, R. D. (1964). Growth and development of cultured plant cells. *Science* **143**, 20–27.

STEWARD, F. C., MOTT, R. L., ISRAEL, H. W., and LUDFORD, P. M. (1970). Proline in the vesicles and sporelings of *Valonia ventricosa* and the concept of cell wall protein, *Nature* (*London*) **225**, 760–762.

STOUT, E. R., and FRITZ, G. J. (1966). Role of oxygen fixation in hydroxyproline biosynthesis by etiolated seedlings. *Plant Physiol.* **41**, 197–202.

THIMANN, K. V., YOCUM, C. S., and HACKETT, D. P. (1954). Terminal oxidases and growth in plant tissues. III. Terminal oxidation in potato tuber tissue. *Arch. Biochem. Biophys.* **53**, 240.

UDENFRIEND, S. (1970). Biosynthesis of hydroxyproline in collagen. *In* "Chemistry and Molecular Biology of the Intercellular Matrix" (E. A. Balazs, ed.), Vol. 1, pp. 370–384. Academic Press, New York.

URIVETZKY, M., FREI, J. M., and MEILMAN, E. (1965). Cell-free collagen biosynthesis and the hydroxylation of sRNA-proline. *Arch. Biochem. Biophys.* **109**, 480–489.

URIVETZKY, M., FREI, J. M., and MEILMAN, E. (1966). Hydroxyprolyl-soluble ribonucleic acid and the biosynthesis of collagen. *Arch. Biochem. Biophys.* **117**, 224–231.

VARNER, J. E. (1964). Gibberellic acid controlled synthesis of α-amylase in barley endosperm. *Plant Physiol.* **39**, 413–415.

WAGNER, R. R., and CYNKIN, M. A. (1971). Glycoprotein biosynthesis. Incorporation of glycosyl groups into endogenous acceptors in a Golgi apparatus-rich fraction of liver. *J. Biol. Chem.* **246**, 143–151.

WHITE, P. R. (1943). "Plant Tissue Culture." Cattell Press, Lancaster, Pennsylvania.

Winslow, R. M., and INGRAM, V. I. (1966). Peptide chain synthesis of human hemoglobins A and A₂. *J. Biol. Chem.* **241**, 1144–1149.

YIP, C. C. (1964). The hydroxylation of proline by horseradish peroxidase. *Biochim. Biophys. Acta* **92**, 395–397.

Lipids and Membrane Structure

Andrew A. Benson

Scripps Institution of Oceanography, University of California at San Diego, La Jolla, California

I. SPECIFICITY OF HYDROPHOBIC ASSOCIATIONS IN LIPOPROTEINS

The specificity of selection of membrane lipids and of their interaction with membrane proteins is remarkable. It is generally overlooked by those studying the structures and properties of lipid bilayers or of membrane proteins. While a few cases of specific selection of hydrocarbon chains might be overlooked as fortuitous, the number of examples of specific fatty acid or other hydrocarbon chain association in membrane lipoprotein is growing and seems too great to be anything but the result of evolutionary selection and design of membrane protein structure to accommodate selected hydrocarbon chains at specific sites. Examples which are recognized but not clearly understood include smell perception, hydrophobic vitamin, antibiotic and hormone action, pheromones, prostaglandins, and steroids. The recognized rather precise structural requirements for biological activity in these substances suggests that they interact with receptor proteins at hydrophobic sites to induce biologically active lipoprotein conformations essential for the selectivity, sensitivity, and adaptability of the organism.

In the lipid bilayer region of a membrane, there can be little requirement for fatty chain specificity beyond those for appropriate fluidity and consequent phase transitions. The liquid or paracrystalline arrangement of hydrocarbon chains in bilayer membranes is well documented. In view of the heterogeneous collection of hydrocarbon chains in most membranes there seems little reason to consider their structural requirements as being very specific.

Lipid bilayer structures are relatively sensitive to solvents. Their lipid is easily removed by ether or chloroform extraction. Many biological membrane systems, however, resist simple solvent extraction and require solvation by at least two types of solvent (viz. chloroform:methanol, 2:1, a usual lipid extractant). Sometimes even this fails and the lipids must be released by enzymatic proteolysis or by strong denaturing solvents like formic acid which unfolds the protein and releases bound hydrophobic molecules.

A dramatic example of selective binding of a hydrophobic molecule in a protein is the red protein which Nishimura and Takamatsu (1957) obtained when they extracted chloroplasts with cold 70% acetone. With almost all lipid material removed, there remained a binding site (Ji et al., 1968) for β-carotene which was so specific it induced a shift of absorption maximum from 480 nm to 538 nm. This specificity is akin to that of smell perception, hormone action, or inhibition of electron transport by DDT and DDE (Bowes and Gee, 1971). The interactions of the hydrocarbon chains of the fatty acids of many complex lipids with membrane protein are often nearly as impressive. We consider these interactions to be the basis for much of the structural uniqueness of membrane lipoprotein (Kaplan and Criddle, 1971).

II. AMPHIPATHIC LIPIDS OF MEMBRANES

Almost all the amphipathic, or surface active, lipids of membranes are fatty acid esters of hydrophilic neutral or anionic small molecules. Lecithin, the galactolipids of plants, and the plant sulfolipid (Fig. 1) are

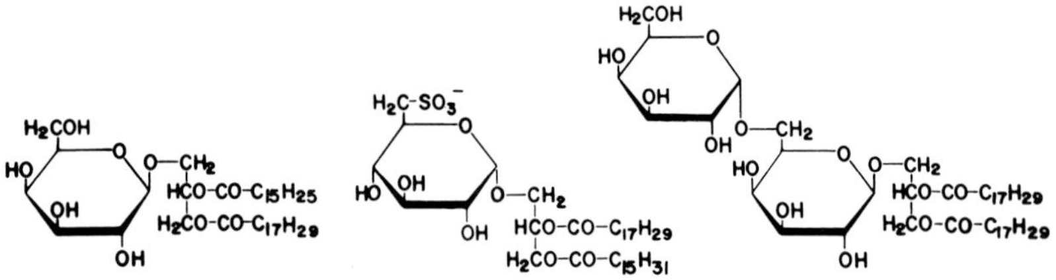

FIG. 1. Glycolipids of chloroplasts: galactosyl diglyceride, sulfoquinovosyl diglyceride, and digalactosyl diglyceride.

typical amphipathic molecules of membrane lipoprotein. Each possesses a very different hydrophilic and/or charged site capable of binding ions or hydrophilic molecules and charged protein surfaces to the membrane. Even anionic lipids like phosphatidylinositol and the sulfolipid which possess similar charge and numbers of hydroxyl groups exhibit unique ion-binding and solvent affinity properties. Their specific association with certain organelles or with lipoprotein components of organelles implies that certain types of membrane function involve hydrophilic groups or certain surface charge and hydrophilicity distribution essential for membrane function. The mitochondria and the salt-secreting glands of both animals and plants are rich in lecithin, suggesting a role for the phosphorylcholine zwitterionic moiety in ion transport.

The galactolipids occur in highest concentrations in chloroplast membranes where evidence presented by Bishop *et al.* (1971) indicates their function on the exterior surfaces of lamellar (thylakoid) membranes. The function of amphipathic lipids in augmenting or altering the properties of membrane proteins is demonstrated by successes in membrane reconstitution (Benson *et al.*, 1970) and lipoprotein activation induced by addition of appropriate lipids or amphipathic detergents.

III. LIPID BILAYER STRUCTURE IN MEMBRANES

The classical "unit membrane" structure of membranes is popularly accepted as a model for molecular relationships within the two black lines of high resolution electron micrographs of osmium-stained sections of membranes. As we shall see later, it is not the only possible model. The lipid bilayer exhibits interesting but not necessarily biological properties. Its state changes, revealed as heat capacity discontinuities in differential thermal analysis studies (Steim, 1970) resemble those observed in some natural membranes. Its nuclear magnetic resonance (NMR) spectra exhibit line broadening indicating semicrystalline restriction of freedom of molecular movement (Cherry *et al.*, 1971). These data indicate a large portion of lipid bilayer structure in some membranes (erythrocyte and *Mycoplasma* membranes). They do not preclude presence of 20% of lipoprotein components of quite another type. No studies by differential thermal analysis of interactions in true lipoproteins have been reported. NMR study of serum lipoproteins indicate restricted molecular movement of methylene groups of the associated lipid chains (Scanu, 1967). A functional role for the unit membrane structure in biological membranes is equivocal, both quantitatively and qualitatively.

The primary evidence for lipid bilayer structure in membranes stems from the frequent occurrence of high lipid:protein ratios in membranes. The amphipathic lipid appears to function as "plasticizer" for membrane lipoprotein and the excess must accumulate in the lamellar micellar structure known as the bilayer. Biological evidence of Frye and Edidin (1970) revealed rapid diffusion of cell surface antigens, not conceivable in a tight mosaic of membrane lipoprotein but consistent with diffusion of proteins in a semifluid bilayer membrane. The models suggested by Singer (1971) are based upon this information.

There appears no physical basis for fatty acid specificity or selectivity in the lipid bilayer membrane. Some fatty acid chains pack better than others, but the observations have always been made with homogeneous lipid preparations. Natural membrane lipids, except for those like the *Mycoplasma* (Smith, 1969) membrane, include a wide spectrum of fatty acid chain lengths and degrees of unsaturation. They can only exist in a semiliquid state within the membrane; there can be no specific structural requirement for chain length or degree of unsaturation. The chain length of a fatty acid is about 20 Å. Hence the width of a lipid bilayer membrane would be 40 Å, rather than the 60–80 Å usually observed in osmium-stained sections. The nature of the bilayer and its staining properties has been reviewed by Korn (1969).

IV. LIPID CONTENT OF BIOLOGICAL MEMBRANES

Myelin, the classical "unit membrane" or bilayer-type membrane is about 80% lipid. Its lipids differ dramatically from those of the glial cell from whence it was elaborated (Davison *et al.*, 1966). Its structure therefore need bear little relationship to the structure and biologically functional membrane of the parent cell.

Membranes of halobacteria consist of 20% lipid. This appears to be a minimum lipid content for adequate "plasticizing" of membrane lipoprotein. Other membrane lipid contents range between these, chloroplast and mitochondrial membranes being about 50% lipid and 50% protein. The variation of chloroplast chlorophyll, sulfolipid, and other lipids with culture conditions suggests too, that a certain amount of lipid bilayer or lipid micellar structure exists in these membrane structures. No doubt this is the source of the DTA and NMR observations which appeared to be characteristic of bilayer structure.

Bishop *et al.* (1971) reported 3-fold higher galactolipid contents in "bundle sheath" than in "mesophyll" chloroplasts of maize and sorghum. Since "mesophyll" chloroplasts possess grana formed from appressed lamellar membrane, it was concluded that galactolipids are essential components of lamellar membrane surfaces. In the "bundle-sheath" chloroplasts where membrane surfaces in contact with the aqueous stroma are reduced in area by the aggregation, the number of sites for galactolipid binding are also apparently reduced. It appears, then, that the lipid compositions of membranes may vary as the lipoprotein components aggregate or disperse. Gunning (1970) has reported the ultrastructural details of membrane condensation which suggest involvement of lipoprotein subunits in the process.

The number of lipid molecules associated with a membrane protein was examined by Ji and Benson (1968). They observed, for each of a group of natural lipid components, a binding stoichiometry of 36 lipid hydrocarbon chains per membrane protein (MW 23,000) unit. In the detergent-isolated membrane lipoprotein preparations, this relationship is maintained, some of the lipids being displaced by equal numbers of detergent hydrophobic chains. The number of hydrophobic sites within membrane protein appears to be fixed (Ji, 1968). Experiments to determine the hydrophobic specificity of these sites have yet to be done.

V. SPECIFICITY OF HYDROCARBON CHAINS ASSOCIATED WITH AMPHIPATHIC LIPIDS

There is a mass of specific but yet circumstantial evidence supporting the tenet that the hydrocarbon moieties of the amphipathic lipids are the keys or determinants for interaction with proteins of membranes. They appear to fit hydrophobic spaces within the membrane, having shapes determined by the protein's amino acid sequence and conformation. The availability of lipid chains most suitable for hydrophobic association within these spaces may be the critical requirement for effective function of the membrane lipoprotein. It appears that the more homogeneous is the membrane lipoprotein, the more unique is its lipid composition.

Hydrophobic association holds certain chlorophyll molecules within the chloroplast's lamellar lipoprotein. In this case the phytol chain of the chlorophyll (Fig. 2) acts as the key and the protein as tumbler holding the porphyrin ring in a position essential for effective electron transport from excited chlorophyll to the redox systems of adjacent electron transport proteins. Extraction experiments have indicated that certain lipids, like chlorophyll, are more tightly bound to protein. In the spinach photosynthetic lipoprotein, P-700, analyzed by C. F. Allen and P. Good (personal communication), the monogalactolipid was completely absent whereas the digalactolipid and others of similar solubility remained. The

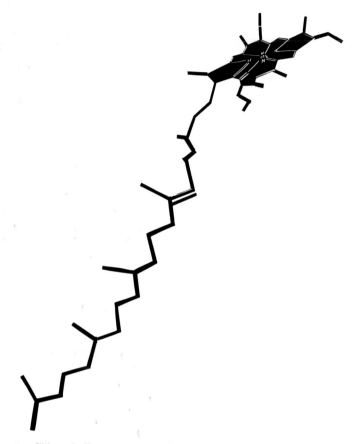

Fig. 2. Chlorophyll, an amphipathic membrane lipid in chloroplasts. The structural similarity of its phytyl chain to polyalanine is clear.

specific differences in fatty acyl components appear to determine the ease of solvent removal of the monogalactosyl diglyceride. Livne and Racker (1969) reported a specific requirement of sulfolipid for activation of coupling factor CF_1 in photophosphorylation. The recent discovery of copious levels of C_{25}-isoprenoid acids in the sulfolipid of many plants (P. J. C. Kuiper and B. Stuiver, personal communication) may provide an explanation for the specific requirement of sulfolipid in this lipoprotein. The requirement of specific lipids such as phosphatidylethanolamine and phosphatidylserine for certain steps in the blood clotting process or for immunochemical interactions may stem from the extraordinary degree of unsaturation in mammalian lipids of these groups. In the marine copepod *Calanus plumchrus*, for example, an ethanolamine sphingomyelin contained 85% 22:6 acid. This was unique among phospholipid components of that organism (S. Patton and R. F. Lee, personal communication). This specific acylation by this extremely oxidation-sensitive acid is a striking example of the remarkable specificity of acylation of amphipathic lipids in membranes.

VI. OXIDATION RESISTANCE OF MEMBRANE LIPIDS

No satisfactory explanation of the general resistance of polyunsaturated membrane lipids to oxidation has been proposed. Presence of carotenoids and other potential antioxidants provides only a partial explanation. In the chloroplast where several atmospheres partial pressure of oxygen are diffusing from the lamellar membranes, where excited states of pigment molecules, free radicals, and peroxides exist in presence of copious amounts linolenic acid esters of the galactolipids, one would expect immediate oxidation as observed when the lipids are separated from their positions in the lamellar membrane. The lipids, however, are remarkably stable, and no oxidation products are observed until they are extracted from the membrane and exposed to air.

In the swimbladders of deep sea fishes, where cholesterol and sphingomyelin accumulate with several hundred atmospheres pressure of oxygen, there exist moderate amounts of arachidonic acid (20:4) and 22:6 acid in a state resistant to oxygen attack (Phleger and Benson, 1971). It is not yet determined whether singlet oxygen can penetrate membrane lipoproteins and attack their polyunsaturated lipids. The problem appears difficult but amenable to investigation (Tinberg and Barber, 1970). The structure of membrane lipoprotein must be consistent with the inaccessibility of its lipids to oxygen.

VII. LIPIDS AND DEVELOPMENT

Cells involved in differentiation produce membranes with structure and function novel to those of the parent. The control of fatty acid synthetic processes and of fatty acylation specificity could well be an important part of this process. Selection of proteins for membrane involvement appears to involve their interaction with certain lipids. Control of lipid availability could thereby lead to control of the type of membrane being elaborated. Although rather little is yet known of variation of lipid components in membranes of selected organelles it is increasingly clear that each membrane lipoprotein may have its own limited spectrum of fatty chain components. The question: Do membrane proteins select their lipid components, or do the lipid components play a role in selecting membrane proteins? already has partial answers. The full story could reveal much about the control of membrane development.

VIII. MEMBRANE MODELS

Assembling molecules and morphology in models stimulates experiment in many areas. Only in this way can the model be tested and discarded or revised. Present concepts of membrane molecular structure include aspects of both the unit membrane micellar structure and the lipoprotein mosaic structure. The latter concept was described by Wallach and Gordon (1968), Scanu (1967), Glaser *et al.* (1970), and Vanderkooi and Green (1970). The conclusions reached by Singer (1971) bring the strength of much contemporary information on membrane protein structure to bear on the subject. His concept of amphiphathic proteins in membranes leaves little said of the specific protein–lipid interactions discussed in this paper. We propose that the lipoprotein molecules floating in the lipid bilayer include lipids as well. The model shown in Fig. 3, described by Singer (1971) as the "lipid-globular protein mosaic model," represents a diffusible amphiphathic membrane protein in a lipid bilayer. In Fig. 4 the floating unit is a lipoprotein possessing lipids with hydrophobic lipid chains specified by the internal structure of the apolipoprotein. A system of this sort meets most of the recognized requirements of membrane structure and can have properties consistent with observed physical properties. The successful reconstitution of many membrane lipoproteins having observable properties or function can hardly involve formation of lipid

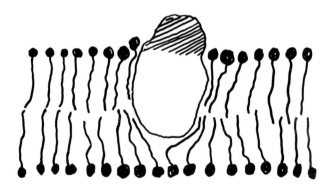

Fig. 3. Lipid-globular protein mosaic model for a membrane.

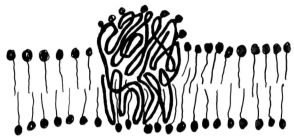

Fig. 4. Amphipathic lipoprotein membrane unit in lipid bilayer membrane.

bilayer structures. Only the lipoprotein molecules themselves could result from reaggregation of lipid and protein. These would be the major sources of enzymatic or biological function and may or may not require reformation of the lipid bilayer which acts as a "plasticizer" or supporting medium for functional lipoproteins.

Membrane models must be consistent with morphological factors observed by ultrastructural techniques. The structure of chloroplast membranes studied by Arntzen et al. (1969) shows promise of being interpretable. The valuable attributes of freeze-etch technology has contributed importantly to the feasibility of understanding molecular arrangement in membranes (Branton, 1971).

REFERENCES

ARNTZEN, C. J., DILLEY, R. A., and CRANE, F. L. (1969). A comparison of chloroplast membrane surfaces visualized by freeze-etch and negative staining techniques; and ultrastructure characterization of membrane fractions obtained from digitonin-treated spinach chloroplasts. *J. Cell Biol.* **43**, 16–31.

BENSON, A. A., GEE, R. W., JR, T.-H., and BOWES, G. W. (1970). Lipid-protein interactions in chloroplast lamellar membrane as bases for reconstitution and biosynthesis. *In* "Autonomy and Biogenesis of Mitochondria and Chloroplasts", (N. K. Boardman, A. W. Linnane, and R. M. Smillie, eds.), pp. 18–26. North-Holland Publ., Amsterdam.

BISHOP, D. G., ANDERSEN, K. S., and SMILLIE, R. M. (1971). The distribution of galactolipids in mesophyll and bundle sheath chloroplasts of maize and sorghum. *Biochim. Biophys. Acta* **231**, 412–414.

BOWES, G. W., and GEE, R. W. (1971). Inhibition of photosynthetic electron transport by DDT and DDE. *Bioenergetics* **2**, 47–60.

BRANTON, D. (1971). Freeze-etching studies of membrane structure. *Phil. Trans. Roy. Soc. London, Ser. B* **261**, 133–138.

CHERRY, J. B., HSU, K., and CHAPMAN, D. (1971). Absorption spectra of chlorophyll in biomolecular membranes. *Biochem. Biophys. Res. Commun.* **43**, 351–358.

DAVISON, A. N., CUZNER, M. L., BANIK, N. L., and OXBERRY, J. (1966). Myelinogenesis in the rat brain. *Nature (London)* **212**, 1373–1374.

FRYE, L. D., and EDIDIN, M. (1970). The rapid intermixing of cell surface antigens of the formation of mouse-human heterokaryons. *J. Cell Sci.* **7**, 319–335.

GLASER, M., SIMPKINS, H., SINGER, S. J., SHEETZ, M., and CHAN, S. I. (1970). On the interaction of lipids and proteins in the red blood cell membrane. *Proc. Nat. Acad. Sci. U.S.* **65**, 721–728.

GUNNING, B. E. S. (1970). Lateral fusion of membranes in bacteroid-containing cells of leguminous root nodules. *J. Cell Sci.* **7**, 307–317.

JI, T.-H. (1968). The structure of chloroplast lamellar membrane. Doctoral Dissertation, University of California, San Diego.

JI, T.-H., and BENSON, A. A. (1968). Association of lipids and proteins in chloroplast membranes. *Biochim. Biophys. Acta* **150**, 686–693.

JI, T.-H., HESS, J. L., and BENSON, A. A. (1968). Studies on chloroplast membrane structure. I. Association of pigments with chloroplast lamellar protein. *Biochim. Biophys. Acta* **150**, 676–685.

KAPLAN, D. M., and CRIDDLE, R. S. (1971). Membrane structural proteins. *Physiol. Rev.* **51**, 249–272.

KORN, E. D. (1969). Cell membranes: Structure and synthesis. *Annu. Rev. Biochem.* **38**, 263–288.

LIVNE, A., and RACKER, E. (1969). Partial resolution of the enzymes catalyzing photophosphorylation. V. Interaction of coupling factor I from chloroplasts with ribonucleic acid and lipids. *J. Biol. Chem.* **244**, 1332–1338.

NISHIMURA, M., and TAKAMATSU, K. (1957). A carotene-protein complex isolated from green leaves. *Nature (London)* **180**, 699–700.

PHLEGER, C. F., and BENSON, A. A. (1971). Cholesterol and hyperbaric oxygen in swimbladders of deep sea fishes. *Nature (London)* **230**, 122.

SCANU, A. (1967). Binding of human serum high density lipoprotein apoprotein with aqueous dispersions of phospholipids. *J. Biol. Chem.* **242**, 711–719.

SINGER, S. J. (1971). The molecular organization of biological membranes. *In* "Structure and Function of Biological Membranes" (L. I. Rothfield, ed.). Academic Press, New York.

SMITH, P. F. (1969). The role of lipids in membrane transport in *Mycoplasma laidlawii. Lipids* **4**, 331–336.

STEIM, J. M. (1970). Thermal phase transitions in biomembranes. *In* "Liquid Crystals and Ordered Fluids," pp. 1–13. Plenum, New York.

TINBERG, H. M., and Barber, A. A. (1970). Studies on vitamin E action: Peroxidation inhibition in structural protein,—lipid micelle complexes derived from rat liver microsomal membranes. *J. Nutr.* **100**, 413–418.

VANDERKOOI, G., and GREEN, D. E. (1970). Biological membrane structure. I. The protein crystal model for membranes. *Proc. Nat. Acad. Sci. U.S.* **66**, 615–621.

WALLACH, D. F. H., and GORDON, A. (1968). Lipid protein interactions in cellular membranes. *Fed. Proc., Fed. Amer. Soc. Exp. Biol.* **27**, 1263–1268.

III. Organization and Expression
of Genetic Information

Organization of DNA and Proteins in Mammalian Chromosomes

Elton Stubblefield

University of Texas at Houston, Department of Biology, Section of Cell Biology, M. D. Anderson Hospital and Tumor Institute, Houston, Texas

I. INTRODUCTION

After almost a century of study, the arrangement of the genes in eukaryotic chromosomes remains an enigma. The methods of genetics have clearly demonstrated that the genes are arranged in linear sequence along the chromosomes, and recently molecular biology has made it quite clear that these genes are composed of the DNA molecules found in chromosomes. Beyond this, however, progress has been slow, mainly for technical reasons. Most chromosomes are too small to be studied in detail with the light microscope, and on the other hand, in the electron microscope most preparations have revealed only a bewildering tangle of fibers. Only in the last decade have cell synchrony techniques provided sufficient amounts of metaphase cells for biochemical isolation procedures to be brought to bear on the problem. In the last five years a variety of approaches have been tried to isolate metaphase chromosomes, and only in recent months have we really learned enough about isolated chromosomes to compare and evaluate the different procedures.

A large number of indirect studies have accumulated over the years which tell us something of the way mammalian chromosomes are organized. We know how newly replicated DNA strands are segregated to daughter and granddaughter cells (Taylor *et al.*, 1957; Deaven and Stubblefield, 1969). We know that chromosome arms can be broken and rejoined rather readily, and that certain rules govern the way the broken ends can be rejoined (Taylor, 1958; Brewen and Peacock, 1969). We even know something of the rates and sequences of DNA replication in individual chromosomes in a few cases (Stubblefield and Gay, 1970). All such evidence must eventually be considered before our understanding of chromosome structure will be complete. However, we must also look directly at chromosomes to see how they are organized and then use the indirect data to help us interpret what we see.

II. OBSERVATIONS AND INTERPRETATION

One of the fundamental units of chromosome structure is the nucleohistone fiber. In the chromosome shown in Fig. 1, this fiber has a diameter of about 250–300 Å. A large fraction of the DNA in the chromosome is found in this form combined with the chromosomal proteins. According to the studies of DuPraw and Bahr (1969), each unit length of 250 Å chromatin fiber contains a minimum of about 50 units of molecular DNA (packing ratio = 1:50). How many parallel DNA molecules are contained in each 250 Å fiber has not been clearly determined; DuPraw and Bahr (1969) and Lampert and Lampert (1970) both think a single DNA molecule is contained in each fiber as a coiled coil. However, since the chromosome is a replicated structure, it is also reasonable to suppose that each 250 Å chromatin fiber might contain two DNA molecules.

Upon treating isolated chromosomes like the one shown in Fig. 1 with 6.0 M urea, all the histone and much of the nonhistone protein is removed; however, the DNA is retained if it was not degraded during isolation by cellular DNase. The fibers of the chromosome are now found to be only 70–100 Å in diameter, as in Fig. 2. Although it is impossible to quantitate the fiber length from such a micrograph, it appears that there are more 70 Å fibers than there were 250 Å fibers in the untreated chromosome. The treated chromosome is somewhat swollen in appearance, but not more than about twice its original size.

Two models for the 250 Å fiber are depicted in Fig. 3. One contains a single DNA thread coiled into a 70 Å fiber which is again coiled into a 250 Å fiber. In the other model two DNA molecules are coiled together in the analogous way (Ris and Chandler, 1963). In both cases the packing ratio can be the same, about 1:50. However, in the first case the conversion of a 250 Å fiber into a 70 Å fiber results in a fiber twice as long as in the other case. Comparison of Figs. 1 and 2, where dissociation into 70 Å fibers was accompanied by an apparent increase in the number of fibers but a minimum of swelling in chromosome size, might lead us to favor the two-stranded model for the 250 Å chromatin fiber. However, theoretical considerations favor the model shown in Fig. 3a (Stubblefield, 1973).

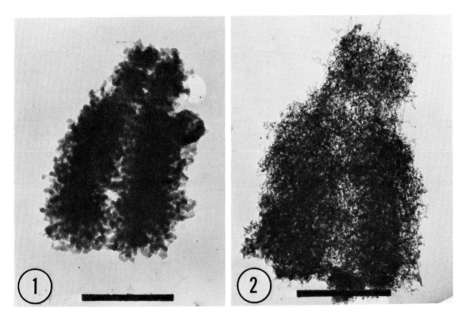

FIG. 1. Chinese hamster metaphase chromosome No. 7 or 8 isolated at Ph 6.5 by the procedure described by Stubblefield and Wray (1971). For details see the legend to Fig. 7. Chromosomes isolated at pH 10.5 have an identical appearance to the one shown here. Anderson critical point dried. Bar = 1 μm.

FIG. 2. Chromosome like the one shown in Fig. 1 isolated at pH 10.5 and then treated for 1 minute with 6.0 M urea before dehydration and drying by the Anderson critical point procedure. Whereas the chromatin fibers were about 250 Å in diameter in Fig. 1, after urea treatment they are about 70 Å in diameter. Bar = 1 μm.

Since removal of the histones by acid treatment (Fig. 4) results in the conversion from 250 Å to 70 Å fibers, it would seem that one or more of the histones are involved in maintaining the chromatin fiber in this state. In agreement with such a role for these molecules is the finding by Kobayashi et al. (1970) that all the major histone fractions are uniformly distributed among the chromosomes of Chinese hamster cells. Our own studies show this also (Fig. 5). In both studies chromosomes were separated by velocity sedimentation on sucrose gradients according to size. In the case of our own experiment, we had expected in the fractions containing B chromosomes an enrichment of any histones involved in the inactivation of the X and Y chromosomes, which comprise up to one-third of these fractions. No such enrichment was detected. This evidence

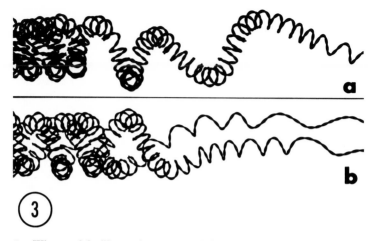

FIG. 3. Wire models illustrating two possible fiber organizations for 250 Å chromatin fibers. In (a) a single 20 Å DNA molecule is coiled into a 70 Å diameter helix which is in turn coiled into the 250 Å fiber. In (b) two such molecules are cohelically wound to form the 250 Å fiber. Either model can achieve packing ratios in excess of 1:50, but converting the 250 Å fiber to 70 Å fibers as in Fig. 2 results in a doubling of the fiber number according to the model in (b).

tends to rule out those theories which propose histones as the control molecules responsible for the late-replication and genetic inactivation of the sex chromosomes in somatic cells. Presumably nonhistone proteins might play this role, as might also a subtle modification of the histones in these chromosomes, such as phosphorylation or acetylation.

Somewhat better resolution of the 250 Å chromatin fiber can be obtained in negative stained preparations of chromosomes. Figure 6 is an example of a chromosome preparation stained with 1% uranyl acetate. In many cases strands about 200 Å in diameter appear to be made of coiled 70 Å fibers, but no clues as to the arrangement of DNA and protein are readily obtained by such an approach.

The length of 250 Å fiber contained in a chromosome is likewise impossible to measure directly in electron micrographs because of the tangled appearance of the chromosomes. However, knowledge of the total DNA content of a metaphase cell, the fraction of the total genome contained in a particular chromosome, and the ratio of the dry mass of a length of chromatin fiber compared to the whole chromosome (measured by electron absorption and scatter) allowed DuPraw and Bahr (1969) to calculate the packing ratio discussed earlier. A chromosome like the one shown in Fig. 1 contains about 3% of the total DNA (15×10^{-12} gm) in a Chinese hamster metaphase cell, or about 0.4×10^{-12} gm. If this amount

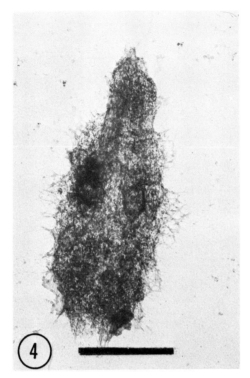

FIG. 4. Chromosome isolated at pH 10.5 and then extracted with 0.2 N HCl for 1 minutes immediately before dehydration and drying by the Anderson critical point method. Removal of the histones by this procedure converts the chromatin fibers to 70 Å. Bar = 1 μm.

of DNA were a single continuous DNA duplex, it would be a fiber about 20 Å in diameter and about 15 cm long. How such a length of DNA is packaged into a structure only 3 μm long is an intriguing mystery. If DuPraw's estimate of a packing ratio of 1:50 is correct, then we may calculate the length of 250 Å chromatin fiber in the chromosome in Fig. 1 to be 3.0 mm. However, since the metaphase chromosome is a double structure, each chromatid would contain only 1.5 mm of chromatin, a fiber about 500 times the chromatid length.

How is such a length of chromatin organized into the chromatid structure? The simplest hypothesis supposes that each chromatid is composed of a single 250 Å fiber coiled and folded into the observed structure. There is considerable evidence consistent with this view, but none of it rules out all other chromosome models. On the other hand, much recent

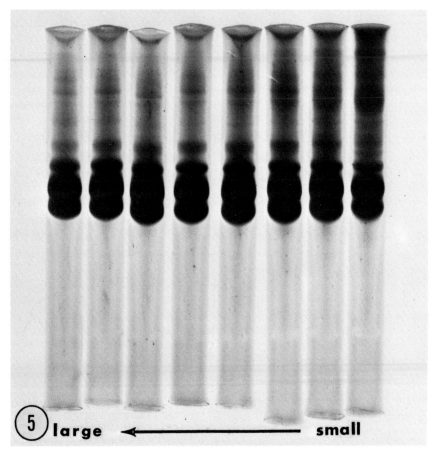

FIG. 5. Acrylamide gel electrophoresis of histones extracted from chromosomes isolated at pH 6.5 after they were fractionated according to size on a 10 to 40% linear sucrose gradient. The small chromosomes on the right were contaminated by cellular debris resulting in additional bands other than the major histones. The histones were extracted in 0.2 N HCl and fractionated and stained according to the procedure described by Wray and Stubblefield (1970).

evidence seems to argue against this simple view, as will soon become apparent.

When chromosomes are isolated at pH 10.5 from Chinese hamster metaphase cells, the activity of cellular DNase is effectively inhibited by the alkaline conditions, and the DNA molecules recovered from the isolated chromosomes are the same size distribution as those found in metaphase cells lysed directly on an alkaline sucrose gradient. Figure 7 shows velocity sedimentation profiles for the DNA from metaphase

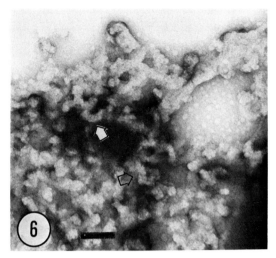

FIG. 6.   250 Å chromatin fibers negatively stained with 1% uranyl acetate. The 70 Å component fibers are visible at the arrows. Bar = 0.1 μm.

cells (a), from chromosomes isolated at pH 10.5 (b) and from chromosomes isolated at pH 6.5 (c). The effects of DNase activity in the latter case are apparent. In the other cases the molecular weights of the denatured DNA molecules range from about 10 million to 200 million daltons. These molecules therefore range upward to about 200 μm in length. Over 1000 such molecules must be somehow assembled to produce the chromosome seen in Fig. 1. Mammalian chromosomal subunits of this size have been observed in several other laboratories (Lett *et al.*, 1970; Painter *et al.*, 1966). Taylor (1968) found that the replication rate of the DNA in mammalian cells was about 1 or 2 μm per minute and estimated the length of a replicating DNA unit (replicon) to be 180 to 360 μm long, allowing 3 hours for its replication. On the other hand, the direct measurement of replicon length in purified DNA by an autoradiographic technique (Huberman and Riggs, 1968) gave an average value of only 60 μm per replicon, and at least several of these were joined end-to-end in many cases. It therefore seems logical to propose that the replicating units of a chromosome correspond to the distribution of DNA lengths which we and others have observed in sucrose gradients and that these units are joined end-to-end into much longer strands by linkages that are destroyed by alkali (Lett *et al.*, 1970). Such a linkage could consist simply of an overlap region where one polynucleotide chain of one molecule is base-paired to a complementary region of the other molecule. Such

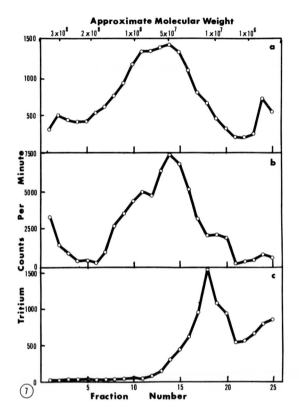

FIG. 7. Constant velocity alkaline sucrose gradient profiles of DNA molecules recovered from metaphase cells (a), chromosomes isolated at pH 10.5 (b), and chromosomes isolated at pH 6.5 (c). In (a) metaphase cells were lysed directly on the gradient, while in (b) and (c) the metaphase chromosomes were isolated from the cells before lysis on the alkaline gradient. In (c) the chromosomes were isolated exactly as described by Stubblefield and Wray (1971) at pH 6.5 in a solution containing 1 M hexylene glycol, 5×10^{-4} M CaCl$_2$, and 1×10^{-4} M PIPES buffer. In (b) the procedure was the same, but the isolation solution contained 1 M hexylene glycol, 2×10^{-3} M CaCl$_2$, and 1×10^{-3} M CAPS buffer at pH 10.5. The top of the gradient is to the right in each case. Note the degradation of the DNA in (c) at pH 6.5. In (b) the molecular sizes range from about 10 million up to about 200 million daltons.

an assembly could presumably produce a continuous fiber of enormous length, and thus determine the linear order of genes in a chromosome, and yet still break into reproducible distributions of subunits upon alkaline denaturation. We shall return to the question of the number of such polymolecules per chromatid after we have examined the architectural design of the chromosome.

The DNA of eukaryotes is interspersed with families of DNA segments having identical or nearly identical base sequences, the repetitive DNA's (Britten and Kohne, 1968). In mammalian cells approximately one-third of the DNA is of this type, with probably many different families present in each species. In the laboratory mouse, for example, about one-third of the repetitive DNA (10% of the total DNA) is all in one large family, the mouse satellite DNA, which was shown by Pardue and Gall (1970) to occupy a position near the centromere in each autosome and X chromosome. The position in the chromosomes of the other repetitive DNA's is undetermined as yet. This particular centromeric repetitive DNA is not found in any other species of mouse, although it appears that many, if not most mammalian chromosomes contain a centromeric repetitive DNA (Hsu and Arrighi, 1971). That there should be so much repetitive DNA in a genome is quite surprising. It now seems certain that large aggregates of this DNA have something to do with chromosomal structure or function at the centromere, and it seems quite probable that the rest of it is important to chromosomal structure and function elsewhere in the chromosome. In reannealing of repetitive DNA's *in situ*, the most frequently observed sites of these DNA's in chromosomes are the centromere and the telomeres, but other sites are apparently scattered along the chromosome arms (Pardue and Gall, 1970).

From their studies, Britten and Kohne (1970) insist that the repeated sequences of DNA occur every few microns throughout the DNA. Such a distribution might suggest a role as a spacer between cistrons, and a similar role as a component of the junction between replicons is also possible. We will return to this idea later.

In a recent publication (Stubblefield and Wray, 1971), we presented evidence from an electron microscope study of isolated Chinese hamster chromosomes which led to the following conclusions about the morphology of the metaphase chromatid: (1) each chromatid is a double structure consisting of two identical halves; (2) each half-chromatid has an axial component (core), a fiber about 500 Å in diameter with many loops of chromatin fibers 250 Å in diameter (epichromatin) attached along its length; (3) the chromatin loops are readily sheared off the core if care is not taken during the isolation procedure; (4) the core fiber can be opened up to reveal that it is a pair of ribbons about 4000 Å wide by 50 Å thick; (5) the core contains about one-fourth of the total DNA in the chromosome. Figure 8 shows a partly sheared chromosome, revealing the core component in each chromatid. In Fig. 9 the double ribbon nature of a core fiber is demonstrated. The model which we proposed in that study is shown in Fig. 10.

After we discovered that chromosomes isolated at neutral or acid pH

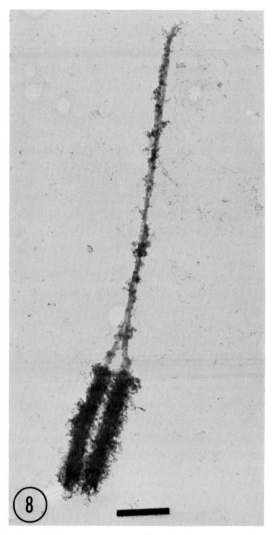

FIG. 8. Partly sheared chromosome revealing a core structure in each chromatid. Bar = 1 μm. Anderson critical point dried.

all suffered DNase breakage of the DNA, we developed the procedure for isolating chromosomes at alkaline pH (Wray *et al.*, 1973) where the DNase activity is inhibited. Treatment of chromosomes isolated at pH 6.5 with 6.0 *M* urea (Fig. 11) resulted in a loss of the 250 Å chromatin fibers (epichromatin) leaving behind the intact chromosome core (Stubblefield and Wray, 1971). However, 6.0 *M* urea treatment of chromosomes

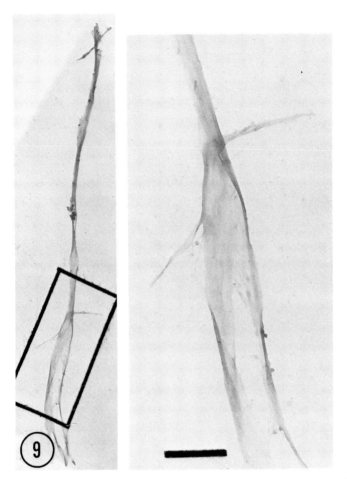

FIG. 9. Isolated chromosome core ribbon over 20 μm long. The enlargement demonstrates that each core fiber contains 2 ribbons face-to-face. Bar = 1 μm.

isolated at pH 10.5 (Fig. 2) did not remove the epichromatin, so it would appear that the DNase activity at pH 6.5 made this effect possible. Presumably the DNase had nicked the DNA while the metaphase cells were swelling in the isolation buffer; when the proteins were later removed from the isolated chromosomes by the urea treatment much of the DNA of the epichromatin fraction simply fell off. This would, of course, require a double-strand break of the DNA where it was attached to the core. That the whole chromosome did not simply disintegrate as a result of the urea treatment is significant.

Fig. 10. Chromosome model proposed by Stubblefield and Wray (1971). The upper left arm is intact, while the lower left arm is pulled out to show its helical nature. Bits of membrane are frequently attached to the telomere. Further extension of the upper right arm reveals the two core fibers, which are shown in the lower right arm with the epichromatin removed. Each core fiber can be unwound to reveal two ribbons about 50 Å thick by 0.4 μm wide. The core fiber would be much longer than shown here.

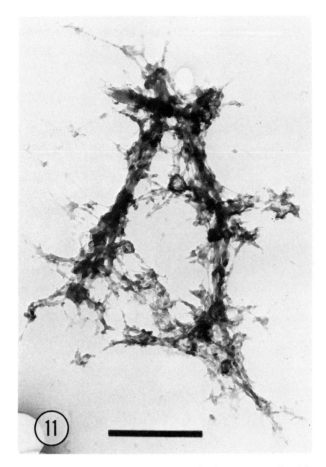

FIG. 11. Chromsome isolated at pH 6.5 and then extracted with 6 M urea. Because of DNase activity much of the epichromatin was removed, leaving the core structures. Compare with Fig. 2. Anderson critical point dried. Bar = 1 μm.

How is the epichromatin attached to the core ribbon? In our earlier report (Stubblefield and Wray, 1971), we had erroneously concluded that ionic bonding was involved; we now know that the two fractions are covalently linked. The nuclease effect allowing the loss of the epichromatin may be what Ohnuki (1968) was describing in his treatment of chromosomes to demonstrate the coiled chromonemata (cores?) in human chromosomes. Likewise the procedures recently developed for *in situ* annealing of radioactive nucleic acids directly into the DNA of chromosomes (Pardue and Gall, 1970) and for staining chromosomal heterochromatin (Hsu and Arrighi, 1971) may also rely on this effect. In fact, in the latter instance it is known that considerable DNA is lost after the DNA is alkali-denatured (T. C. Hsu and F. E. Arrighi, personal communication). That *in situ* annealing of nucleic acids works at all is a good indication that the chromosome core is rich in repetitive DNA, since it is only this fraction of DNA that can be hybridized in these preparations.

On the basis of these considerations it seems quite possible that the 250 Å epichromatin fibers are attached to the core ribbon through complementary binding of the bases in repetitive DNA. If these repetitive DNA's are also at the ends of replicons, then we might also suspect that they may in some way control the replication of the large aggregates of chromatin which are duplicated together in the chromosome, e.g., early- or late-labeling regions. Such an arrangement would require that about half of the repetitive DNA be contained in the core ribbons and the remainder distributed throughout the chromatin at replicon ends (excluding similar DNA's used as spacers between cistrons). This arrangement also makes quite interesting possibilities for interpreting the process of meiotic exchange and the structure of the synaptinemal complex.

The doubleness of the chromatid requires the presence of two chromatin fibers 250 Å in diameter attached at intervals along the two core fibers in each chromatid. If each 250 Å fiber contains a single DNA molecule, then in effect the half-chromatid would be uninemic in the sense of containing a single long DNA molecule from one end to the other.

The major problem with the chromosome model presented here is to understand the meaning of genetic mutation where multiple copies of a gene are present. Some mechanism must operate in meiosis to correlate all four copies of each gene and correct any random errors that might have arisen. Of course, such a mechanism would lend itself to genetic stability and the faithful transmission of the genetic message from generation to generation.

ACKNOWLEDGMENTS

The author wishes to gratefully acknowledge the capable technical assistance of Miss Nita Rivoire. This work was supported in part by a research grant from the National Science Foundation (GB-16250).

REFERENCES

BREWEN, J. G., and PEACOCK, W. J. (1969). Restricted rejoining of chromosomal subunits in aberration formation: A test for subunit dissimilarity. *Proc. Nat. Acad. Sci. U.S.* **62,** 389–394.

BRITTEN, R. J., and KOHNE, D. E. (1968). Repeated sequences in DNA. *Science* **161**, 529–540.

BRITTEN, R. J., and KOHNE, D. E. (1970). Repeated segments of DNA. *Sci. Amer.* **222**, 24–31.

DEAVEN, L. L., and STUBBLEFIELD, E. (1969). Segregation of chromosomal DNA in Chinese hamster fibroblasts *in vitro. Exp. Cell Res.* **55**, 132–135.

DuPRAW, E. J., and BAHR, G. F. (1969). The arrangement of DNA in human chromosomes, as investigated by quantitative electron microscopy. *Acta Cytol.* **13**, 188–205.

HSU, T. C., and ARRIGHI, F. E. (1971). Distribution of constitutive heterochromatin in mammalian chromosomes. *Chromosoma* **34**, 243–253.

HUBERMAN, J. A., and RIGGS, A. D. (1968). On the mechanism of DNA replication in mammalian chromosomes. *J. Mol. Biol.* **32**, 327–341.

KOBAYASHI, S., MAIO, J. J., and SCHILDKRAUT, C. L. (1970). Histones of fractionated metaphase chromosomes of Chinese hamster cells. *Fed. Proc., Fed. Amer. Soc. Exp. Biol.* **29**, 730A.

LAMPERT, F., and LAMPERT, P. (1970). Ultrastructure of the human chromosome fiber. *Humangenetik* **11**, 9–17.

LETT, J. T., KLUCIS, E. S., and SUN, C. (1970). On the size of the DNA in the mammalian chromosome. Structural subunits. *Biophys. J.* **10**, 277–292.

OHNUKI, Y. (1968). Structure of chromosomes. I. Morphological studies of the spiral structure of human chromosomes. *Chromosoma* **25**, 402–428.

PAINTER, R. B., JERMANY, D. A., and RASMUSSEN, R. E. (1966). A method to determine the number of DNA replicating units in cultured mammalian cells. *J. Mol. Biol.* **17**, 47–56

PARDUE, M. L., and GALL, J. G. (1970). Chromosomal localization of mouse satellite DNA. *Science* **168**, 1356–1358.

RIS, H., and CHANDLER, B. L. (1963). The Ultrastructure of Genetic Systems in Prokaryotes and Eukaryotes. *Cold Spring Harbor Symp. Quant. Biol.* **28**, 1–8.

STUBBLEFIELD, E. (1973). The Structure of Mammalian Chromosomes. *Internat. Rev. Cytol.* **35**, 1–60.

STUBBLEFIELD, E., and GAY, M. L. (1970). Quantitative tritium autoradiography of mammalian chromosomes. II. The kinetics of DNA synthesis in individual chromosomes of Chinese hamster fibroblasts. *Chromosoma* **31**, 79–90.

STUBBLEFIELD, E., and WRAY, W. (1971). Architecture of the Chinese hamster metaphase chromosomes. *Chromosoma* **32**, 262–294.

TAYLOR, J. H. (1958). Sister chromatid exchanges in tritium labelled chromosomes. *Genetics* **43**, 515–529.

TAYLOR, J. H. (1968). Rates of chain growth and units of replication in DNA of mammalian chromosomes. *J. Mol. Biol.* **31**, 579–594.

TAYLOR, J. H., WOODS, P. S., and HUGHES, W. L. (1957). The organization and duplication of chromosomes as revealed by autoradiographic studies using tritium labeled thymidine. *Proc. Nat. Acad. Sci. U.S.* **43**, 122–128.

WRAY, W., and STUBBLEFIELD, E. (1970). A highly sensitive procedure for detection of histones in polyacrylamide gels. *Anal. Biochem.* **38**, 454–460.

WRAY, W., HUMPHREY, R., and STUBBLEFIELD, E. (1973). Mammalian metaphase chromosomes with high molecular weight DNA isolated at pH 10.5. *Nature (London) New Biol.* **238**, 237–238.

The Multiple Relations of tRNA to Metabolic Control

ROBERT M. BOCK

Department of Molecular Biology, University of Wisconsin, Madison, Wisconsin

As we learn more details about balanced regulation of metabolism, development of organisms and their adaptation to changes in the environment, we see the regulated state to be the result of a complex continuum of interactions involving every catalyst, structural component, and metabolite of a living tissue. At some stage or under some state of stress, each component of the cell takes on importance in the maintenance of regulation during growth and development.

We know of many roles for proteins. As catalysts they change activity in response to pH, substrate concentrations, presence of allosteric factors, and degree of assembly of the subunits in polymeric enzymes or multienzyme complexes. As structural building blocks of organelles and tissues, the role of proteins is also dynamic and regulated. Today no one would be foolhardy enough to present a paper entitled "The Role of Proteins in Metabolic Regulation." The myriad of roles is well accepted, and scientists now pursue specific details at the identified levels of regulation.

I feel that we are now on the verge of moving to a similar level of sophistication with respect to transfer RNA. Thus, if I am not the last to discuss in a broad general manner the roles of tRNA in metabolic control, at least we can hope that the day will soon come when our colleagues will not attempt to examine these many tRNA roles in a single overview. However, the path leading to this level of sophistication in which the specific roles can be studied is a path of many small steps which wander in apparently unguided directions. I will try in this broad overview to sort out a few of the roles which tRNA does or does not seem to have. I will try to point out some routes for more direct inquiry in search of better understanding of several specific roles.

Perhaps a look at some of the regulatory processes in which proteins participate will better focus our attention on valid questions that could be asked in a symposium entitled "Macromolecules Regulating Growth and Development." We know of numerous cases where the catalytic activity of a single protein varies in response to the immediate reactants and products and to the pH, temperature, and concentration of cofactors within the cell. In addition, many of these enzymes show more complex variability of their activity in response to the environment. For example, allosteric interactions occur with metabolites other than the immediate reactant and product, or occasionally the reactants and products themselves have an additional allosteric effect. These allosteric effects result in kinetic behavior more complex than predictable by simple mass action relationships. The reaction of transfer RNA with the enzyme aminoacyl tRNA synthetase has all the above regulatory aspects including the observation that transfer RNA itself behaves as an allosteric activator and stabilizer of the synthetase enzyme. Whether the ribosome, that multienzyme workbench upon which tRNA performs its most important role, is also subject to allosteric regulatory phenomena involving tRNA has not yet been testable with the state of refinement of our *in vitro* protein synthesis systems. Are these among the regulatory roles we seek to document? I feel they should not be ignored.

Proteins are found to differ markedly when states of organ development or differing tissues are compared to each other. The numerous stages of fetal and adult hemoglobins are a particularly well documented example of this phenomenon. The discovery of multiple forms of enzymes, the isozyme phenomenon, led workers to compare many tissues, so there is now a wealth of comparative data about the number and types of a particular enzyme in various stages of development and in differentiated tissues. In spite of this rich array of data correlating isozymes to developmental or metabolic state, it is rare indeed that a positive cause and effect relationship between the enzyme under discussion and the maintenance or production of a particular state of development can be satisfactorily established. Returning to the hemoglobin case, we note that the fetal and adult stages of this protein often exhibit responses of oxygen binding to pH, to carbon dioxide, and to metabolites which can be incorporated into a speculative model proposing these phenomena as highly amenable to this stage of development of the organism. However, this highly regulated, intricately developing system of oxygen transport is ordinarily viewed as a result, not the cause, of development. Even the famous Hill effect and Bohr effect which influence oxygen transport in hemoglobin remained as circumstantial evidence of control phenomena until mutant hemoglobins were found to differ in these properties with interesting resultant physiological changes.

Weighing the evidence for regulatory involvement of any single protein will tame our wild hopes for discovery of dramatic new regulatory phenomena. After decades of study we have only the poorly understood hormone proteins, the repressor proteins and the sigma factor of RNA polymerase as proteins so intimately involved in regulation that their role could be causal, and not simply an effect. The histones remain an enigma. I feel that as we learn more about the details of growth and development, many other macromolecules whose variations have been correlated with development will be found to play a substantial role in the maintenance of a given differentiative state or the progress to the next state of differentiation. The macromolecule may be essential as part of a morphological assembly which imparts the characteristic recognized as that state of differentiation. It may be that the macromolecule is a catalyst or an essential component of a catalytic process which produces a new metabolite or a different level of a more ubiquitous metabolite and thus sustains the state of differentiation or allows development to continue. It is only in rare cases that one macromolecule appears to play a dominant specific and active role as the key to development. Even induction and repression are frequently small percentage changes in the rate of synthesis of an enzyme. Enzymes showing large inductive and repressive changes have been selected for many studies, hence one gets the impression that large changes are typical.

When transfer RNA was first discovered there was great enthusiasm that it would play a key and central role in metabolic regulation. After all, here was the molecule which was the Rosetta Stone of biology. It was able to participate in the recognition of specific amino acid metabolites and could also recognize nucleic acids. Here was a strong candidate for the specificity of readout of genetic information. Studies of metabolic behavior induced or repressed by amino acids soon included tests of the role of tRNA in these processes. The effect of inhibitors and the study of mutant systems with defective aminoacyl tRNA synthetases implicated deacylated tRNA as a possible repressor of general RNA synthesis. More recent studies have focused attention on the magic spot, guanosine tetraphosphate as a key metabolite in this control of RNA synthesis (Cashel and Kalbacker, 1970). Yet there is circumstantial evidence that since tRNA has a terminal guanosine phosphate and undergoes additional guanylation reactions, it may be involved in the production of this regulatory guanosine tetraphosphate. Although the role of tRNA in the general control of RNA synthesis (and relaxation of control by certain mutants) has been studied more than any other regulatory phenomena involving tRNA, we are still without a conclusive answer whether tRNA plays an important role or not.

Let us look briefly at some of the *in vitro* behavior of transfer RNA systems and then return to the *in vivo* correlations involving it in regulation. The allosteric effects of transfer RNA on the aminoacyl tRNA synthetase were documented in detail with *in vitro* systems. The *in vitro* requirement for ATP to maintain the terminal adenosine active for accepting an amino acid is easily demonstrated, as is the rapid turnover of this terminus in living cells. Experiments have not documented, however, whether there is a causal effect wherein loss of this terminal adenosine retards the rate of protein synthesis when a cell is under a stress of limited ATP availability.

The existence of transfer RNA molecules which carry the same amino acid but read differing codons has led to the speculation that there may be a regulatory phenomenon which involves differential rates of translation of messages depending on whether or not they contain codons for which an adequate supply of the appropriate tRNA is present. This process would, of course, demand some way that the transfer RNA could be changed in relative abundance or in efficiency for reading particular codons depending on the metabolic state or the stage of differentiation of the cell. It is well known that transfer RNA for a particular amino acid can be found on several different chromosomes of the yeast cell (Mortimer, 1969). Differential activity of these chromosomes during aerobic or anaerobic growth, sporulation, germination and other developmental states could change the supply of transfer RNA if the chromosomes were differentially activated.

Direct correlations of codon selective translation have been found in *in vitro* studies of isolated transfer RNA. Gefter and Russell (1969) observed that transfer RNA which had not been modified to convert the adenosine adjacent to the anticodon into the hormone cytokinin was much less effective in translating coded messages than after that modification was completed. Since transfer RNAs containing the cytokinin hormone have been found only in those tRNAs which read code letters beginning with U, it has been proposed that this could be an effective means of selection among the various messages available. More recently, Ohashi *et al.* (1970) have observed that the uracil in the position which reads the third letter of the codon triplet can, in certain tRNA's, undergo thiolation and additional modification with a resultant change in ability to read the possible codons. This immediately generates a regulatory model in which the degree of modification at this position would adapt this transfer RNA for preferentially reading messages depending on whether they were rich in one set of codons or another. No *in vivo* evidence for such a variability in codon reading has been obtained for this case although Sueoka and Kano-Sueoka (1970) have reported evidence that the leucine tRNA produced after phage infection of *E. coli* is better fitted to read the codons abundant in the phage genome in preference to the host genetic messages.

We have tested the prediction that in the presence of adequate cytokinin a tissue will synthesize protein with more of the amino acids specified by these tRNA's. While the short-term rates of amino acid incorporation into tissues were in general accord with this prediction, there was a definite exception in that histidine incorporation increased as much or more with added cytokinin even though it does not occupy the U codon row (M. Kaminek and R. M. Bock, unpublished results, 1971). Also, the total accumulation of amino acids into bulk protein did not show the correlation which was found in short-term uptake studies. If one of the many antisuppressor mutants in yeast is found to prevent the cytokinin modification of tRNA, it will indeed be of great importance to workers who hope to understand the role of this modification.

In vitro tests have been able to show that certain transfer RNAs are definitely not involved in any regulatory role which demands the matching of a particular amino acid to a certain codon. At least these tRNA's were not involved under the growth conditions in which those particular cells were cultured. For example, deletion mutants of certain T4 bacteria phage were found to have the eight tRNA genes carried by this bacteria phage totally deleted without loss of the ability of the phage to develop in an apparently normal manner (W. McClain, personal communication, 1971).

Squires and Carbon (1971) found that one of the genes for glycine tRNA in *E. coli* specified a transfer RNA which read the codons GGA and GGG. When this gene mutated and became defective, the cells were not viable. They were restored to viability by mutation in a gene specifying a glycine transfer RNA which normally read the codons GGU and GGC. It mutated to read GGA and GGG. Fortunately, the cell contained multiple copies of the genes for reading GGU and GGC, and so it could spare one to take over the function of the defective GGA, GGG tRNA gene. The evidence was strong that the mutation occurred only within the anticodon; therefore any peculiar properties of tRNA modification, interaction with activating enzymes, or other metabolic roles which correlated certain codons to an amino acid are not compatible with this observed replaceability of one gene for another.

The *in vivo* correlations between the number and properties of tRNA species and the state of metabolism or differentiation are so numerous that I will not bore you with an attempt to compile a comprehensive list. The amount and type of tRNA has been found to vary between mitochondria and cytoplasm within a given cell line, between red cells and muscle cells, between liver cells with and without estrogen treatment, between spores and vegetative cells of a given organism and within a host bacterium before and after bacteriophage development. The amount and types of tRNA's vary between embryonic tissue and the adult and between certain tumor lines and the presumed organ of origin. However, all these observations are correlations and are not yet documented cases of cause and effect. It appears that either a reconstructed *in vitro* system or genetic evidence for strong and essential dependence on certain of these tRNA's for differentiation will be needed before any causal arguments can be supported.

Studies on the histidine operon have revealed the most definite information about the role of tRNA in metabolic regulation. The original speculations that tRNA could serve to modulate the relative amount of gene product from various cistrons within one operon (Ames and Hartman, 1963) or that tRNA could control the initiation of different messages by varying the level of tRNA for the initiator sequences of messages (Stent, 1964) have had no convincing demonstrations after many years of study.

Some very strong evidence of other roles has appeared. Even though there has been no evidence of variable types of initiator tRNA's as the metabolism changes, the state of formylation of the normal chain initiator has been shown to keep an operon tightly under operator control at limited formyl tRNA levels, but not when formyl methionyl tRNA was abundantly present (Gerberich *et al.*, 1967). Lewis and Ames (1971) have used bacterial cells with a mutant tRNA, or mutant histidyl tRNA synthetase, and even a presumably mutant tRNA modifying enzyme to document that histidyl tRNA in complex with its activating enzyme is essential for the repression of the histidine biosynthesis operon and might be the, repressor itself. One mutant was found to have an undermodified tRNA functional for protein synthesis but not for repressor function. However, consider the histidine operon as an example of the complexity of metabolic regulation. This operon is definitely known to be under control of the operator–repressor system; it can be shifted out of this control by high levels of formylmethionyl tRNA; it can be shown to be subject to translation control after the message is produced. The enzymes resultant from the operon are subject to variations in their catalytic effectiveness as the cellular environment changes.

Additional evidence exists that tRNA could be an inducer or repressor. When Wilson (1970) examined cells in which both ethionine and methionine could acylate tRNA, both were inducers of 5-adenosylmethionine (SAM) reductase. However, in mutants that could not attach ethionine to tRNA, the ethionine did not induce SAM reductase. This led him to speculate that met-tRNA was the real inducer. It could also be that tRNA^{met} was the repressor and that methionine plus the synthetase were able to remove or inactivate the repressor. Other modes of action, including translation where some ethionine incorporation was needed, are still possible explanations.

While the genetic attack on the role of tRNA in metabolic regulation appears to me to be one of the most powerful and essential for documenting the variety of roles played, it has not been applied in a direct selective search for tRNA's altered in their regulatory capacity. The many cases of suppressor transfer RNA which have been found are clearly not of central interest to metabolic regulation. If the minor transfer RNA which was altered in codon reading to become a suppressor had been involved in a key regulatory step, it could not have been spared for this mutation. In yeast, tyrosine suppressor tRNA's from eight different genetic loci have been allowed to become suppressors. If all these were topaz suppressors, that is, functional for reading the normal tyrosine codons and also the ochre codon (Table 1), then they could serve their regulatory role even though mutated. If so, this family is of primary importance in order to find nonfunctional mutants of the suppressor and thus test whether each of these eight tyrosine genes can be spared from any or all developmental and metabolic functions of the cell. There are known in yeast some suppressors which are lethal when in the homozygous diploid state (Cox, 1971). This lethality could result if they were the only tRNA's capable of reading a certain codon or if they were essential for a particular developmental phenomenon. The former possibility is open to direct test by analysis of the tRNA complement of homozygous and heterozygous strains.

Gorman and Gorman (1971) have selected a revertant to a missense-suppressing yeast strain with properties that imply that some tRNA modifications necessary for mitochondrial function may be unneeded for cytoplasmic functions. The reverting mutation was in a gene other than the tRNA sequence itself, and the loss of function of the missense-suppressing tRNA was accompanied by a loss of function of mitochondria. Mutagenesis and selection of those cells which lost function of one (or more) suppressor tRNA's and simultaneously lose a developmental function (mitochondrial function, sporulation, etc.) would yield evidence on the relation of tRNA modification to development.

In general the study of mutant transfer RNA's has not been designed to reveal whether the tRNA's had regulatory roles. In an unpredictable event, Jacobson (1971) discovered that tyrosine tRNA was involved in the inactivation of a mutant tryptophan pyrrolase enzyme and that when the tyrosine tRNA mutated, homozygous mutant cells (with no wild-type tRNA present) no longer caused inactivation of the enzyme. Jacobson

TABLE 1

PATTERNS OF CODON RECOGNITION BY tRNA's[a,b]

1st letter	U	C	2nd letter A	G	3rd letter
U	Phe	Ser	Tyr	Cys	U
	Phe	Ser	Tyr	Cys	C
	Leu	Ser	Ochre/Topaz	Umber	A
	Leu	Ser	Amber	Trp	G
C	Leu	Pro	His	Arg	U
	Leu	Pro	His	Arg	C
	Leu	Pro	Gln	Arg	A
	Leu	Pro	Gln	Arg	G
A	Ile	Thr	Asn	Ser	U
	Ile	Thr	Asn	Ser	C
	Ile	Thr	Lys	Arg	A
	Met	Thr	Lys	Arg	G
G	Val	Ala	Asp	Gly	U
	Val	Ala	Asp	Gly	C
	Val	Ala	Glu	Gly	A
	Val	Ala	Glu	Gly	G

[a] Adapted from Söll et al. (1967).

[b] The interconnected circles indicate that a single species of tRNA reads one, two, or three triplet codons (○, yeast; ●, Escherichia coli). While the reading of the chain terminator codons UAA and UAG has given rise to the reading patterns called Topaz, Ochre and Amber, the reading pattern of the chain terminator codon UGA has not been systematically named. Since both Opal and Umber have been used as names of suppressors reading this codon, it would be helpful to use the name Opal for the suppressor reading UGU, UGC, and UGA and restrict Umber to the suppressor tRNA, which reads UGA and UGG.

presumes this to be a reflection of some wild-type interaction between tRNA and the enzyme, but there is no evidence cited that this existed nor that the tRNA had any regulatory role in a normal cell. It is, however, at least the third report of a transfer RNA bound to an enzyme apparently unrelated to the tRNA function.

If clever new selective systems are developed for pursuing conditional lethal mutants in tRNA, there is a great opportunity for learning whether transfer RNA's are involved in development or regulation processes. At the present time, there are excellent mapping studies of tRNA genes underway, particularly in yeast (Leupold, 1970; Hawthorne, 1969). By documenting the locus of all of the tRNA genes on the yeast map, one automatically increases the probability that he would recognize the correlation between a developmental defect and the locus of a tRNA gene. This mapping has been aided both by the genetic efforts of the group here and by new techniques for hybridization of labeled tRNA (Schweizer et al., 1969) or ferritin-coupled tRNA (Wu and Davidson, 1971) to the DNA of the cell being mapped. If one concentrates on the suppressor tRNA's which have gained the suppressor function but not lost their normal function, there is a greater probability of finding regulatory mutants. The topaz suppressor, where a normal tyrosine tRNA gains the ability to read the ochre codon, is one such class (Bock, 1967). In addition, there are similar functional suppressors for cysteine transfer RNA and tryptophan transfer RNA (see Table 1).

To explore whether tRNA modification is essential for function or regulation one seeks mutants in the enzymes which modify tRNA. Those mutant genes which cause a loss of suppressor function but are not mapped within the suppressor tRNA locus should be mutant in the modifier enzyme genes. The selective search for such mutants should be done under a variety of metabolic stresses, such as anaerobic, aerobic, sporulation, germination, and high and low temperature growth. Exploitation of such selective systems could lead to such findings as that one or more tRNA modifications are needed only for respiration, not for fermentation. The system would then produce a means of isolating the modified and unmodified tRNA so that its in vitro functions could be compared.

The tRNA modification which has captured the particular interest of my laboratory is one that converts the adenosine adjacent to the anticodon in certain tRNA's to the plant hormone cytokinin. It is now known that animals, plants, and bacteria with the exception of some strains of mycoplasma all have cytokinin in their transfer RNA molecules. The presence of this modified base has been shown to change the in vitro capability of the transfer RNA for recognizing its codon. It does not significantly alter the ability of the tRNA to accept the amino acid from the activating enzyme. Many tissues are known to be able to produce this modification in the tRNA without any dependence on an added hormone. In the case of the hormone-dependent tissue, Burrows et al. (1971) have shown that added cytokinin hormone does find its way into tRNA although, simultaneously, alternate routes exist for synthesizing this modification in the tRNA. It is not known whether this added cytokinin finds its way to all the tRNA's in the U row of codons which contain cytokinin or to only selected species. It is also not known whether the presence in tRNA is directly correlated with any or all of the many hormonal functions expressed by cytokinin.

We clearly need mutants in tRNA modifying enzymes (preferentially temperature-sensitive mutants) to better understand the role of such tRNA modification. In the meanwhile, we are examining the correlation of the amounts of various tRNA species with cytokinin treatment of dependent cells and will attempt to find whether *in vitro* efficiency of the tRNA correlates with its cytokinin content and/or with the level of cytokinin in the growth medium. Hecht and his colleagues (1971) have recently synthesized a potent antagonist of cytokinin, and we will attempt to see whether any tRNA functions are altered by the antagonist.

REFERENCES

ALLENDE, C. C., CHAIMOVICH, H., GATICA, M., and ALLENDE, J. E. (1970). Aminoacyl tRNA synthetases. *J. Biol. Chem.* **245**, 93–101.

AMES, B. N., and HARTMAN, P. E. (1963). The histidine operon. *Cold Spring Harbor Symp. Quant. Biol.* **28**, 349–356.

BOCK, R. M. (1967). Prediction of a topaz suppressor tRNA. *J. Theor. Biol.* **16**, 438–439.

BURROWS, W. J., SKOOG, F., and LEONARD, N. J. (1971). Isolation and identification of cytokinins located in tRNA of tobacco callus grown in the presence of 6-benzyl-aminopurine. *Biochemistry* **10**, 2189–2194.

CASHEL, M., and KALBACHER, B. (1970). The control of RNA synthesis in *E. coli*. *J. Biol. Chem.* **245**, 2309–2318.

COX, B. S. (1971). Recessive lethal suppressors in yeast. *Heredity* **26**, 211–232.

GEFTER, M. L., and RUSSELL, R. L. (1969). Role of modifications in tyrosine transfer RNA: A modified base affecting ribosome binding. *J. Mol. Biol.* **39**, 145–157.

GERBERICH, M., KOVACH, J. S., and GOLDBERGER, R. F. (1967). Chain initiation in a polycistronic message: Sequential versus simultaneous derepression of the enzymes for histidine biosynthesis in *Salmonella typhimurium*. *Proc. Nat. Acad. Sci. U.S.* **57**, 1857–1864.

GORMAN, J. A., and GORMAN, J. (1971). Genetic analysis of a gene required for the expression of allele-specific missense suppression in *Saccharomyces cerevisiae*. *Genetics* **67**, 337–352.

HAWTHORNE, D. C. (1969). Identification of nonsense codons in yeast. *J. Mol. Biol.* **43**, 71–75.

HECHT, S. M., BOCK, R. M., SCHMITZ, R., SKOOG, F., and LEONARD, N. J. (1971). Cytokinins: Development of a potent antagonist. *Proc. Nat. Acad. Sci. U.S.* **68**, 2608–2610.

JACOBSON, K. B. (1971). Role of an isoacceptor tRNA as an enzyme inhibitor. *Nature (London) New Biol.* **231**, 17–19.

LEUPOLD, U. (1970). Genetic studies on nonsense suppressors in *Schizosaccharomyces pombe*. *Heredity* **25**, 493.

LEWIS, J. A., and AMES, B. N. (1971). Charging of tRNA^His *in vivo* and its relation to repression of the Histidine operon, *Fed. Proc. Fed. Amer. Soc. Exp. Biol.* **30**, 1262.

MORTIMER, R. K. (1969). Genetic redundancy in yeast. *Genetics Suppl.* **61**, 329–334.

OHASHI, Z., SANEYOSHI, M., HARADA, F., HARA, H., and NISHIMURA, S. (1970). Presumed anticodon structure of tRNA^{glu} from *E. coli. Biochem. Biophys. Res. Commun.* **40**, 866–872.

SCHWEIZER, E., MACKECHNIE, C., and HALVORSON, H. O. (1969). The redundancy of ribosomal and transfer RNA genes in *Saccharomyces cerevisiae. J. Mol. Biol.* **40**, 261–277.

SÖLL, D., CHERAYIL, J. D., and BOCK, R. M. (1967). Studies on polynucleotides. *J. Mol. Biol.* **29**, 97–112.

SQUIRES, C., and CARBON, J. (1971). Normal and mutant glycine transfer RNA. *Nature (London) New Biol.* **233**, 274–277.

STENT, G. S. (1964). The operon: On its third anniversary. *Science* **144**, 816–820.

SUEOKA, N., and KANO-SUEOKA, T. (1970). Transfer RNA and cell differentiation. *Progr. Nucl. Acid Res. Mol. Biol.* **10**, 23–55.

WILSON, R. H. (1970). Possible methionine-tRNA involvement in a methionine-regulated enzyme system in *Coprinus lagopus. Heredity* **25**, 490.

WU, M., and DAVIDSON, N. (1971). Physical mapping of tRNA genes using ferritin as an electron opaque label. *Fed. Proc., Fed. Amer. Soc. Exp. Biol.* **30**, 1054.

Total Synthesis of Transfer RNA Genes

H. G. Khorana

*Departments of Biology and Chemistry, Massachusetts Institute of Technology,
Cambridge, Massachusetts*

Methods have been developed for the chemical synthesis of short
deoxypolynucleotides. Using these methods together with polynucleotide
kinase and ligase, the total synthesis of the gene for yeast alanine transfer
RNA has been accomplished. The synthesis of *Escherichia coli* tyrosine
RNA is in progress. Details of our work may be found in the following
references:

Khorana, H. G., Agarwal, K. L., Büclin, H., Caruthers, M. H., Gupta, N. K.,
Kleppe, K., Kumar, A., Ohtsuka, E., Raj Bhandary, U. L., Van de Sande J. H., and
Yamada, T. (1972). CIII. Total synthesis of the structural gene for an alanine
transfer ribonucleic acid from yeast. *J. Mol. Biol.* **72**, 209–217.

Weber, H., and Khorana, H. G. (1972). CIV. Total Synthesis of the Structural
Gene for an Alanine Transfer Ribonucleic Acid from Yeast. Chemical synthesis
of an icosadeoxyribonucleotide corresponding to the nucleotide sequence 21 to
40. *J. Mol. Biol.* **72**, 219–249.

Büclin, H., and Khorana, H. G. (1972). CV. Total synthesis of the structural
gene for an alanine transfer ribonucleic acid from yeast. Chemical synthesis of
an icosadeoxyribonucleotide corresponding to the nucleotide Sequence 31 to 50.
J. Mol. Biol. **72**, 251–288.

Kumar, A., Ohtsuka, E., and Khorana, H. G (1972). CVI. Total synthesis of
the structural gene for an alanine transfer ribonucleic acid from yeast. Synthesis
of two nonanucleotides and a heptanucleotide corresponding to nucleotide sequences
22 to 30, 41 to 49 and 28 to 34. *J. Mol. Biol.* **72**, 289–307.

Ohtsuka, E., Kumar, A., and Khorana, H. G. (1972). CVII. Total synthesis
of the structural gene for an alanine transfer ribonucleic acid from yeast. Synthesis
of a dodecadeoxynucleotide and a hexadeoxynucleotide corresponding to the nucleotide sequence 1 to 12. *J. Mol. Biol.* **72**, 309–27.

Kumar, A., and Khorana, H. G. (1972). CVIII. Total synthesis of the structural
gene for an alanine transfer ribonucleic acid from yeast. Synthesis of an undecade-
oxynucleotide, a decadeoxynucleotide and an octadeoxynucleotide corresponding
to the nucleotide sequences 7 to 27. *J. Mol. Biol.* **72**, 329–349.

Agarwal, K. L., Kumar, A., and Khorana, H. G (1972). CIX. Total synthesis
of the structural gene for an alanine transfer ribonucleic acid from yeast. Synthesis
of a dodecadeoxynucleotide and a decadeoxynucleotide corresponding to the nucleotide sequence 46 to 65. *J. Mol. Biol.* **72**, 351–373.

Caruthers, M. H., Van de Sande, J. H., and Khorana, H. G. (1972). CX. Total synthesis of the structural gene for an alanine transfer of ribonucleic acid from yeast. Synthesis of three decadeoxynucleotides corresponding to the nucleotide sequence 51 to 70. *J. Mol. Biol.* **72,** 375–405.

Caruthers, M. H., and Khorana, H. G. (1972). CXI. Total synthesis of the structural gene for an alanine transfer ribonucleic acid from yeast. Synthesis of a dodecadeoxynucleotide and a heptadeoxynucleotide corresponding to the nucleotide sequence 66 to 77. *J. Mol. Biol.* **72,** 407–426.

Sgaramella, V., and Khorana, H. G. (1972). CXII. Total synthesis of the structural gene for an alanine transfer RNA from yeast. Enzymic joining of the chemically synthesized polydeoxynucleotides to form the DNA duplex representing nucleotide sequence 1 to 20. *J. Mol. Biol.* **72,** 427–444.

Sgaramella, V., Kleppe, K., Terao, T., Gupta, N. K., and Khorana, H. G. (1972). CXIII. Total synthesis of the structural gene for an alanine transfer RNA from yeast. Enzymic joining of the chemically synthesized segments to form the DNA duplex corresponding to the nucleotide sequence 17 to 50. *J. Mol. Biol.* **72,** 445–456.

Van de Sande, J. H., Caruthers, M. H., Sgaramella, V., Yamada, T., and Khorana, H. G. (1972). CXIV. Total synthesis of the structural gene for an alanine transfer RNA from yeast. Enzymic joining of the chemically synthesized segments to form the DNA duplex corresponding to nucleotide sequence 46 to 77. *J. Mol. Biol.* **72,** 457–474.

Caruthers, M. H., Kleppe, K., Van de Sande, J. H., Sgaramella, V., Agarwal, K. L., Büclin, H., Gupta, N. K., Kumar, A., Ohtsuka, E., Raj Bhandary, U. L., Terao, T., Weber, H., Yamada, T., and Khorana, H. G. (1972). CXV. Total synthesis of the structural gene for an alanine transfer RNA from yeast. Enzymic joining to form the total DNA duplex. *J. Mol. Biol.* **72,** 475–492.

Biosynthesis of Bacterial Ribosomes

MASAYASU NOMURA

Institute for Enzyme Research, Departments of Biochemistry and Genetics,
University of Wisconsin, Madison, Wisconsin

Studies on the biosynthesis of bacterial ribosomes have been carried out in our laboratory using genetic as well as biochemical approaches. The biochemical approaches aimed at characterization of all the components contained in the ribosomes and elucidation of the structural organization of these complex organelles. In addition, the mechanism of assembly of the ribosomes can be studied *in vitro* when one can reconstitute ribosomes from dissociated molecular components. We first succeeded in reconstituting ribosomes by using 30 S ribosomal subunits from *Escherichia coli;* the 30 S ribosomal subunits were reconstituted from 16 S RNA and a mixture of unseparated 30 S ribosomal proteins (Traub and Nomura, 1968). The reconstituted 30 S particles were shown to be similar to the original 30 S subunits functionally as well as structurally (Traub and Nomura, 1968). These experiments established the concept that the information for the correct assembly of ribosomal particles is contained in the structure of their molecular components, not in nonribosomal factors.

Kinetic studies on the assembly reaction have shown that there is a rate-limiting unimolecular reaction which has a high activation energy (Traub and Nomura, 1969). At lower temperatures only some of the 30 S ribosomal proteins (called RI proteins) interact with 16 S RNA, leading to accumulation of intermediate particles (reconstitution intermediate, or "RI particles"). The RI particles sediment at 21 S, are deficient in several proteins (called S proteins) and do not have any functional activity. From this and other experiments with isolated RI particles, the following reaction scheme has been proposed (Traub and Nomura, 1969):

$$16 \text{ S RNA} \xrightarrow{\text{+RI proteins}} \text{RI particles} \xrightarrow{\text{heating}} \text{RI* particles}$$
$$\xrightarrow{\text{+S proteins}} 30 \text{ S ribosomal subunits}$$

Although we have analyzed the protein composition of the isolated RI particle preparation, the identification of the proteins in the RI particles, as defined by the above assembly scheme, has not been completed.

Next, we separated and purified 30 S ribosomal proteins (Nomura *et al.*, 1969; Ozaki *et al.*, 1969; Held *et al.*, 1973) as did other investigators (Hardy *et al.*, 1969; Kaltschmidt *et al.*, 1967; Craven *et al.*, 1969; Fogel and Sypherd, 1968; Traut *et al.*, 1969; Hindennach *et al.*, 1971), and studied the assembly of 30 S subunits from these purified components. We have found that under the conditions of reconstitution, only seven proteins bind to the RNA and certain others bind only when some of the first seven are bound. The remaining proteins require the presence of proteins in both of the above groups for their binding. In this way, we analyzed the sequence of addition of proteins to the 16 S RNA molecule and constructed an assembly map (Mizushima and Nomura, 1970; Nashimoto *et al.*, 1971).

It is highly probable that the "sequence" described in the map corresponds at least roughly to the temporal sequence of the assembly. In fact, proteins found in the isolated RI particle preparation are all in the "early" part of the sequence, and proteins that are not found in the RI particle preparation are in the "late" part in the map (Nashimoto *et al.*, 1971; Held and Nomura, 1973).

It is clear that the assembly reaction is highly cooperative in the sense that the binding of many proteins depends on the presence of other proteins. The simplest interpretation of the interdependence is that this reflects the topological relationships among ribosomal proteins in the ribosome structure (Mizushima and Nomura, 1970). Other approaches such as those using bifunctional reagents to cross-link neighboring proteins may be used to find out whether this interpretation is correct.

We have recently succeeded in the reconstitution of *Bacillus stearothermophilus* 50 S subunits from their dissociated molecular components (Nomura and Erdmann, 1970; Fahnestock *et al.*, 1973). In comparison with the reconstitution of *E. coli* 30 S subunits, reconstitution of *B. stearothermophilus* 50 S subunits requires higher temperatures and the rate of reconstitution is much slower even at the optimum temperature (60°C). Although 50 S ribosomal proteins from *B. stearothermophilus* have not yet been individually purified, and the reconstitution system is still not refined, the present system may be useful for studies on the mechanism of assembly of 50 S subunits, as well as on the functional role of individual components. For example, using this reconstitution system, we have established that 5 S RNA is essential for reconstitution of the functional 50 S particles (Erdmann *et al.*, 1971).

As noted above, the assembly reaction in the present system is too slow to account for the efficiency of the 50 S ribosome assembly reaction *in vivo*. Perhaps, our *in vitro* conditions are not fully optimized. Alternatively, the *in vivo* assembly may be much faster because it uses "nascent" submethylated precursor 23 S RNA's, which are growing on the DNA template, instead of mature 23 S RNA. However, it should be emphasized that reconstitution of 30 S subunits from the same organism is much faster [by a factor of about 300 at 50°C (Nomura and Erdmann, 1970)] under identical *in vitro* conditions and that this reconstitution is performed using mature 16 S RNA. Such considerations suggest that the assembly of 50 S subunits is facilitated *in vivo* by some special mechanism not involved in the 30 S assembly *in vivo*.

The mechanism of assembly of ribosomes in intact bacterial cells has been studied using cold-sensitive mutants of *E. coli*, which cannot assemble ribosomes at low temperatures (Tai *et al.*, 1969). Our studies have shown that all mutational alterations discovered so far which inhibit 30 S ribosome assembly also inhibit 50 S assembly, whereas there are several mutations that abolish 50 S assembly without affecting 30 S assembly (Guthrie *et al.*, 1969; Nashimoto and Nomura, 1970). One mutant studied in detail is a spontaneous spectinomycin-resistant mutant that cannot grow at 20°C. This mutant is unable to assemble both 30 S and 50 S subunits at 20°C in complex media, and accumulates two kinds of incomplete particles, 21 S and "30 S" particles, which are related to 30 S and 50 S subunits, respectively. A single mutation, causing an alteration in one of the 30 S ribosomal proteins, P4, is responsible for these phenotypes (Nashimoto *et al.*, 1971; Nashimoto and Nomura, 1970). These results suggest that assembly of 50 S subunits *in vivo* is dependent on simultaneous assembly of 30 S subunits, whereas the assembly of 30 S subunits is independent of 50 S assembly. The precise mechanism by which the mutation affecting a 30 S ribosomal protein affects both 30 S and 50 S assembly is not known.

Both RNA and proteins in the 21 S particles accumulated by the mutant have been analyzed. The RNA (precursor 16 S RNA) is slightly larger than mature 16 S RNA (Nashimoto and Nomura, 1970; Lowry and Dahlberg, 1971). The RNA is also submethylated or not methylated at all (Lowry and Dahlberg, 1971). About 11 out of 21 ribosomal proteins in the 30 S subunits are present in the 21 S particles, but the remaining 30 S ribosomal proteins are absent (Nashimoto *et al.*, 1971). All the proteins found in the 21 S particles are in the early part of the *in vitro* assembly map. It is concluded that the order of addition of proteins during the *in vivo* 30 S assembly is similar to the order of addition of proteins during the *in vitro* reconstitution.

Although the *in vitro* reconstitution reaction has provided much useful information on the structure, function, and assembly process of ribosomal particles, it is clear that the *in vivo* assembly process is not exactly the same as the *in vitro* process. One clear difference is the use of precursor ribosomal RNA *in vivo*. It is probable that ribosomal proteins may start to bind to ribosomal RNA while the latter is being synthesized on DNA template. It is equally possible that the interacting ribosomal proteins are also in precursor forms which are different from mature ribosomal proteins used for reconstitution. Thus, the assembly process should perhaps be studied *in vitro* using DNA preparations coding for ribosomal RNA and ribosomal proteins together with all the components necessary for transcription, translation, and posttranscriptional (or translational) modifications. Complete reproduction *in vitro* of all the assembly events that occur *in vivo*, although it may be difficult, should be the eventual goal. It should then become possible to study *in vitro* any factor regulating biosynthesis of the ribosomes, a process which is important for the growth and/or development of any organism.

ACKNOWLEDGMENTS

The author wishes to thank his previous as well as present associates who have contributed to the work described here. Our work has been supported by Grant GM-15422 from the National Institutes of Health and by Grant GB-31086X1 from the National Science Foundation.

REFERENCES

CRAVEN, G. R., VOYNOW, P., HARDY, S. J. S., and KURLAND, C. G. (1969). The ribosomal proteins of *Escherichia coli*. II. Chemical and physical characterization of the 30 S ribosomal proteins. *Biochemistry* **8**, 2906–2915.

ERDMANN, V. A., FAHNESTOCK, S., HIGO, K., and NOMURA, M. (1971). Role of 5 S RNA in functions of the 50 S ribosomal subunits. *Proc. Nat. Acad. Sci. U.S.* **68**, 2932–2936.

FAHNESTOCK, S., ERDMANN, V., and NOMURA, M. (1973). Reconstitution of 50 S ribosomal subunits from protein-free ribonucleic acid. *Biochemistry* **12**, 220–224.

FOGEL, S., and SYPHERD, P. S. (1968). Chemical basis for heterogeneity of ribosomal proteins. *Proc. Nat. Acad. Sci. U.S.* **59**, 1329–1336.

GUTHRIE, C., NASHIMOTO, H., and NOMURA, M. (1969). Structure and function of *E. coli* ribosomes. VIII. Cold-sensitive mutants defective in ribosome assembly. *Proc. Nat. Acad. Sci. U.S.* **63**, 384–391.

HARDY, S. J. S., KURLAND, C. G., VOYNOW, P., and MORA, G. (1969). The ribosomal proteins of *Escherichia coli*. I. Purification of the 30 S ribosomal proteins. *Biochemistry* **8**, 2897–2905.

HELD, W., and NOMURA, M. (1973). Rate-determining step in the reconstitution of *Escherichia coli* 30 S ribosomal subunits. *Biochemistry* **12**, 3273–3281.

HELD, W., MIZUSHIMA, S., and NOMURA, M. (1973). Reconstitution of *Escherichia coli* 30 S ribosomal subunits from purified molecular components. *J. Biol. Chem.* **248,** 5720–5730.

HINDENNACH, I., STÖFFLER, G., and WITTMAN, H. G. (1971). Ribosomal proteins. Isolation of the proteins from 30 S ribosomal subunits of *Escherichia coli. Eur. J. Biochem.* **23,** 7–11.

KALTSCHMIDT, E., DZIONARA, M., DONNER, D., and WITTMANN, H. G. (1967). Ribosomal proteins. I. Isolation, amino acid composition, molecular weights and peptide mapping of proteins from *E. coli* ribosomes. *Mol. Gen. Genet.* **100,** 364–373.

LOWRY, C. V., and DAHLBERG, J. (1971). Structural differences between the 16 S ribosomal RNA of *E. coli* and its precursor. *Nature (London) New Biol.* **232,** 52–54.

MIZUSHIMA, S., and NOMURA, M. (1970). Assembly mapping of 30 S ribosomal proteins from *E. coli. Nature (London)* **226,** 1214–1218.

NASHIMOTO, H., and NOMURA, M. (1970). Structure and function of bacterial ribosomes. XI. Dependence of 50 S ribosomal assembly on simultaneous assembly of 30 S subunits. *Proc. Nat. Acad. Sci. U.S.* **67,** 1440–1447.

NASHIMOTO, H., HELD, W., KALTSCHMIDT, E., and NOMURA, M. (1971). Structure and function of bacterial ribosomes. XII. Accumulation of 21 S particles by some cold-sensitive mutants of *Escherichia coli. J. Mol. Biol.* **62,** 121–138.

NOMURA, M., and ERDMANN, V. A. (1970). Reconstruction of 50 S ribosomal subunits from dissociated molecular components. *Nature (London)* **228,** 744–748.

NOMURA, M., MIZUSHIMA, S., OZAKI, M., TRAUB, P., and LOWRY, C. V. (1969). Structure and function of ribosomes and their molecular components. *Cold Spring Harbor Symp. Quant. Biol.* **34,** 49–61.

OZAKI, M., MIZUSHIMA, S., and NOMURA, M. (1969). Identification and functional characterization of the protein controlled by the streptomycin-resistant locus in *E. coli. Nature (London)* **222,** 333–339.

TAI, P., KESSLER, D. P., and INGRAHAM, J. (1969). Cold-sensitive mutations in *Salmonella typhimurium* which affect ribosome synthesis. *J. Bacteriol.* **97,** 1298–1304.

TRAUB, P., and NOMURA, M. (1968). Structure and function of *E. coli* ribosomes. V. Reconstitution of functionally active 30 S particles from RNA and proteins. *Proc. Nat. Acad. Sci. U.S.* **59,** 777–784.

TRAUB, P., and NOMURA, M. (1969). Structure and function of *Escherichia coli* ribosomes. VI. Mechanism of assembly of 30 S ribosomes studied *in vitro. J. Mol. Biol.* **40,** 391–413.

TRAUT, R. R., DELIUS, H., AHMAD-SADEK, C., BICKLE, T. A., PEARSON, P., and TISSIÈRES, A. (1969). Ribosomal proteins of *Escherichia coli.* Stoichiometry and implications for ribosome structure. *Cold Spring Harbor Symp. Quant. Biol.* **34,** 25–38.

The Structural Basis of Selective Gene Expression in Eukaryotes

BRIAN J. McCARTHY[1] AND MICHEL JANOWSKI[2]

Departments of Biochemistry and Genetics, University of Washington, Seattle, Washington

I. INTRODUCTION

An understanding of the mechanism of gene expression in eukaryotic cells would be greatly facilitated by the ability to separate active and inactive segments of chromatin. Such a separation can be achived by shearing and differential centrifugation (Frenster *et al.*, 1963; Frenster, 1969). The same general approach has been used with success to demonstrate the localization of highly redundant satellite DNA sequences in condensed segments of chromatin (Yunis and Yasmineh, 1971; Duerksen and McCarthy, 1971).

Since the size of mammalian and many other eukaryotic genomes is so large compared to that of bacteria, it might be supposed that the fraction of transcribed DNA in any cell is quite small. Certainly a few percent of the mammalian genome would suffice to specify all known enzymes and structural proteins. However, this argument is complicated by the existence of giant heterogeneous nuclear RNA (HnRNA) assumed to be precursor to messenger RNA. If only a minor part of each HnRNA

---

[1] Present address: Department of Biochemistry and Biophysics, University of California, San Francisco, California.

[2] Present address: Laboratories du C.E.N./S.C.K., Mol, Belgium.

molecule becomes transformed to messenger, it is conceivable that the fraction of the genome transcribed is large even in specialized cells where only a few messengers are translated. Measurement of the percent of the genome transcribed may not therefore be useful in distinguishing spacer from informational DNA. Nevertheless, for purposes of effecting an efficient fractionation of chromatin it is important to establish what fraction of the genome is transcribed. In the most general terms, the fraction of the genome transcribed in a single cell is equivalent to the fraction of active chromatin. In practice, this equation may not be so exact, since cell populations are often heterogeneous and different sites in the genome may be transcribed at different rates. Nevertheless, a consideration of the fraction of the genome transcribed is relevant to undertaking chromatin fractionation since the isolation of active chromatin is inherently more difficult in cases where genome expression is quite limited.

In the succeeding pages, we review the facts concerning the extent of transcription in various eukaryotic cells and compare some physical and chemical properties of chromatin isolated from various cells and organs. In addition some data are presented concerning one method of chromatin fractionation that has proved to be useful.

II. THE EXTENT OF TRANSCRIPTION IN EUKARYOTIC CELLS

The transcriptional diversity of a population of RNA molecules may be measured by an excess RNA hybridization reaction. In eukaryotes where the existence of operationally redundant DNA sequences complicates the interpretation of such experiments (Britten and Kohne, 1968; McCarthy and Church, 1970), quantitative estimates can be made only in reference to the unique part of the genome. In general, labeled single-stranded fragments of unique DNA are prepared by annealing and hydroxyapatite fractionation and hybridized with RNA in gross excess for extended periods of time (Gelderman et al., 1970; Davidson and Hough, 1971; Hahn and Laird, 1971; Brown and Church, 1971; Grouse et al., 1972). The same kind of experiment has been conducted by several groups with a variety of eukaryotic cells. Many of these results are tabulated for comparative purposes (Table 1). Included are estimates for two species of bacteria from which it appears that the majority of the genome (80–100%) is expressed assuming that only one DNA strand is transcribed in every region. In eukaryotes, it is apparent that especially for organisms with relatively small genomes, such as the cellular slime mold and *Drosophila melanogaster* the fraction of the genome transcribed may amount to a sizable fraction of the total. It should be emphasized at

TABLE 1

THE EXTENT OF TRANSCPIPTION IN EUKARYOTIC CELLS

	Genome size (daltons)	Percent transcribed	Reference
Dictyostelium discoideum	30×10^9		
Amoebae		15	Firtel (1972)
Total		28	
Drosophila melanogaster	120×10^9		
Cultured cells		15	Turner and Laird (1973)
Embryos		15	Figure 1
Adults		10	
Xenopus laevis	2×10^{12}		
Oocytes		0.6	Davidson and Hough (1971)
Chicken	1×10^{12}		
Red cells		2	McConaughy and McCarthy (1972)
Liver		15	
Mouse	3×10^{12}		
Hepatoma cells		2	Gelderman *et al.* (1971) Hahn and Laird (1971)
Liver, kidney, spleen		4–5	Brown and Church (1971) Grouse *et al.* (1972)
Embryo		8	
Brain		11	

this point that all these estimates are necessarily minimal ones for obvious technical reasons involving the concentration of rarer RNA species and the long incubation times employed (Grouse *et al.*, 1972). In the case of vertebrates, a pattern is apparent in the sense that the more specialized cell types or the more limited mixtures of cells found in tissues, give the lowest values.

One obviously favorable situation is offered by *Drosophila* cells, where about 15% of the DNA is transcribed. Assuming that only one of the two DNA strands is transcribed, this result suggests that some 30% of the chromatin is active. In this case a successful fractionation by sucrose gradient centrifugation has been reported (McCarthy *et al.*, 1973).

III. CHEMICAL AND PHYSICAL PROPERTIES OF CHROMATIN

Most reported analyses of chromatin have shown a remarkably constant chemical composition. This is especially true of histone contents although several authors report differences in the relative amounts of

TABLE 2

ACIDIC AND BASIC PROTEIN CONTENTS OF THE CHROMATINS FROM VARIOUS
MOUSE ORGANS, CORRELATED WITH SPECTRAL RATIOS[a]

Tissue	Basic protein/DNA (μg/μg)	Acidic protein/DNA (μg/μg)	A_{260}/A_{240}	A_{260}/A_{280}
Spleen	0.82	0.33	1.58	1.77
Liver	0.86	0.59	1.46	1.74
Brain	0.87	0.66	1.47	1.73
Kidney	0.87	0.79	1.36	1.68

[a] Basic and acidic proteins were separated by extraction with 0.2 N H_2SO_4. Protein and DNA were quantitated by standard colorimetric methods.

nonhistone proteins (Baserga and Stein, 1971). However to what extent these are nonchromosomal contaminants remains to be shown. Typical analyses for mouse spleen, liver, brain, and kidney chromatin are tabulated (Table 2). Basic protein contents range from 0.82 to 0.87 μg of protein per microgram of DNA. On the other hand, acidic protein contents range from 0.33 to 0.82. In each case the method of chromatin preparation was the same (Bonner et al., 1968) and the chromatin was purified by sedimentation through 1.7 M sucrose in 10 mM Tris buffer, pH 8.0 (Duerksen and McCarthy, 1971).

If these differences in apparent protein content represented genuine firmly associated chromosomal proteins, one might expect a difference in the buoyant density of these four chromatin preparations. For this purpose a sample of each chromatin was fixed with formaldehyde and banded to equilibrium in a CsCl gradient (Brutlag et al., 1969). Despite the fact that buoyant density is a sensitive measure of protein:DNA ratio, all the chromatin samples banded at essentially the same density 1.411 to 1.413 (Fig. 1). Calculations based on the data of Table 2 predicted values ranging from 1.365 for kidney to 1.410 for spleen. From this we conclude that apparent differences in protein content may reflect differential contamination of nonchromosomal proteins.

Again, if the ratio of acidic to basic proteins is substantially different among several chromatin samples, a difference in electrophoretic mobility may occur. Therefore, electrophoresis of the four chromatin samples was compared. All the chromatin samples migrated the same distance after a 3-hour run (Fig. 2). At least two distinct chromatin components appear during the migration. However, the minor moving components have the same ultraviolet spectrum as the main band, and their significance is presently unknown. We have not investigated the nature of the sub-

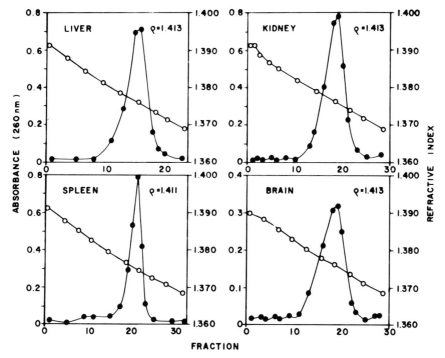

FIG. 1. Buoyant density of chromatin in CsCl. The chromatin preparations were first dialyzed against 0.02 M sodium phosphate buffer (pH 7.0) and then treated for 60 minutes at 0°C with 1.5% formaldehyde. A saturated CsCl solution in 0.02 M sodium phosphate (pH 7.0) and 1.5% formaldehyde was added to give a final density of 1.420 gm/cm³. Centrifugation was carried out for 48 hours at 33,000 rpm in the Spinco SW 39 rotor. Care was taken when reading the refractive indices of the fractions collected through the bottom of the tubes to correct for the presence of phosphate and of formaldehyde in the CsCl solution. The values to be substrated were determined empirically.

stances that accumulate near the cathode. Since there appears to be some correlation between the amount of this material and the acidic protein contents of the electrophoresed chromatins, these may be the nonchromosomal contaminants.

Polyacrylamide gel electrophoresis of the basic chromosomal proteins display the characteristic pattern of histones (Fig. 3). Thus we can conclude that, whatever the meaning of nonhistone chromosomal proteins, the histones do not vary from one kind of chromatin to another either in amount or distribution of the various histone classes.

Based on the data described above it would appear that chromatin has a constant chemical composition regardless of its source (Itzhaki and

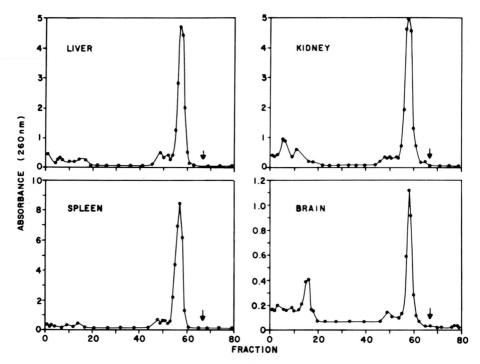

Fig. 2. Zone electrophoresis of chromatin in solution. The method was modified from that of Olivera *et al.* (1964). The apparatus consisted in a LKB No. 8100 Ampholine column containing, from bottom to top, the solutions tabulated below.

Layer No.	NaCl molarity	Tris . HCl (pH 7.6) molarity	% Sucrose	Volume (ml)	Remarks
1	0.1	0.2	60	75	Anode solution
2	0.2	0.02	55	20	
3	0.01	0.001	50	20	
4	0.01	0.001	45–15	320	Linear gradient
5	0.01	0.001	12.5	5	sample
6	0.01	0.001	10	20	
7	0.2	0.02	5	20	
8	0.1	0.2	0	40	Cathode solution

Electrophoresis was carried out at 400 V (which corresponded to 48 mA) for 3 hours. Longer runs resulted in a visible precipitation of the main chromatin band. The column was cooled with ice-cold circulating water throughout the experiment. Approximately 6-ml fractions were collected at the bottom of the column with a peristalic pump.

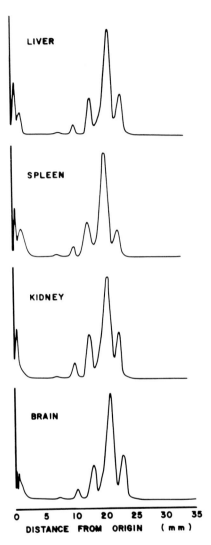

Fig. 3.   Histone content of chromatin from various mouse organs. Histones were prepared and processed for gel electrophoresis on 7.5% polyacrylamide gels as described by Bonner *et al.* (1968).

Cooper, 1973). Nor is there any evidence for heterogeneity or subclasses of chromatin which differ in protein:DNA ratio or charge properties. Thus it would seem that methods for chromatin fractionation must be based on more subtle chemical or physical attributes, which distinguish active and inactive portions of the genome.

IV. THERMAL DENATURATION OF CHROMATIN

It is well established that the thermal denaturation profiles of chromatin extend over a broader range of temperature than those for the DNA. Moreover, the mean thermal denaturation temperature is several degrees higher than for DNA. This presumably results from the stabilization of the DNA double strand by the associated proteins. In accordance with the proposition that various segments of chromatin may be differentially stabilized by proteins in a manner connected with gene regulation, chromatin melting curves often show evidence of multiple components. In particular, it has been suggested that the lower melting components might be representative of genetically active segments (Huang *et al.*, 1964). Indeed, Frenster's experiments show a lower melting temperature in active than in repressed chromatin isolated from the same tissue (Frenster, 1969).

Some examples of thermal denaturation profiles of chromatin are given in Fig. 4. The profiles for mouse spleen, liver, brain, and kidney chromatin are essentially superimposable. This is also true when the same data are plotted on probability coordinates, where several distinct melting components are revealed.

However, complete identity of melting profiles of chromatin from different sources is not a general rule. When chromatin from very different sources is compared, considerable differences in thermal denaturation behavior are evident. This is also illustrated in Fig. 4 through comparisons of mouse liver, sea urchin blastula, and avian embryo erythrocyte chromatin. Especially for the third case most of the chromatin denatured at a higher temperature consistent with the view that, in these highly specialized cells, most of the DNA is transcriptionally inert and more intimately associated with chromosomal proteins.

The purpose of experiments to be subsequently described was to demonstrate that the small proportion of red cell chromatin which does display transcriptional activity is specifically low melting compared to the remainder. In order to carry out this test, it was necessary to adapt the differential melting behavior as a preparative tool for the thermal fractionation of chromatin. Since thermal elution of DNA from hydorxyapatite is an efficient means of fractionating DNA based upon melting temperature (Miyazawa and Thomas, 1965), the same principle was applied to chromatin.

Thermal elution from hydroxyapatite was compared to thermal denaturation measured optically for both chick erythrocyte and mouse liver chromatin (Fig. 5). The chromatin was absorbed at 60°C with 100%

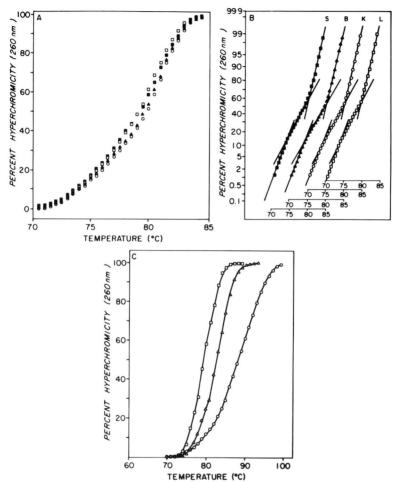

FIG. 4. Thermal denaturation profiles of various chromatin samples. The chromatin samples were dissolved at an approximate absorbancy of 1.0 at 260 nm in 0.01 M Tris · HCl (pH 8.0). They were heated with a temperature increment of 1°C per minute in a Beckman spectrophotometer, using a Haake circulating water bath. The hyperchromic shift at 260 nm was recorded with a Gilford multisample absorbance recorder. (A, C) ■, Mouse spleen; □, mouse liver; ▲, mouse brain; △, sea urchin embryo; ○, chick embryo erythrocyte. (B) The profiles for mouse brain (B), kidney (K), liver (L), and spleen (S) are displayed on probability coordinates.

efficiency and eluted in the range of 78–100°C. A small fraction remaining at 100°C was recovered after washing with 8 M urea. As is the case for DNA, the temperature of elution varied slightly according to the batch of hydroxyapatite and the supplier. Batches from Clarkson gave consis-

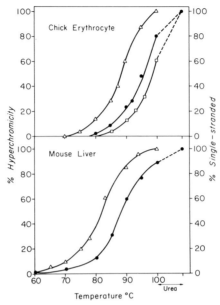

Fɪɢ. 5. Thermal denaturation profiles of chick erythrocyte and mouse liver chromatin measured spectrophotometrically or by hydroxyapatite chromatography. Sheared chromatin was heated at 50 μg/ml in 0.025 *M* phosphate buffer (pH 6.8), and the hyperchromicity was measured as a function of temperature in a spectrophotometer at 260 nm in 1-cm pathlength cuvettes (△). Sheared chromatin was also dissociated on a Bio-Rad (●) or Clarkson (□) hydroxyapatite column in 0.12 *M* phosphate buffer, and the amount of single stranded material was monitored as a function of temperature. Elution of any material remaining at 100° was accomplished by washing the column with 8 *M* urea–0.24 phosphate–0.01 *M* EDTA.

tently higher T_m's than did those sold by Bio-Rad. Nevertheless in both cases the elution profiles paralleled the optical denaturation profiles measured in the spectrophotometer.

The DNA eluted from the column appeared to be single-stranded associated with some protein. However, some chromosomal protein appears to be lost in the procedure.

V. CHARACTERIZATION OF DNA IN FRACTIONATED CHROMATIN

If the fractionation based upon differential thermal elution has biological significance, then the DNA in the various fractions might differ in base sequence and complementarity with red cell RNA. In particular, we wished to test the possibility that the first few percent eluted at the lowest temperature contained the DNA base sequences transcribed in chick erythrocytes. For this purpose chromatin was prepared from chick embryos exposed to [³H]thymidine from day 8 to day 15 of incubation. The red cell preparation was contaminated by less than 0.1% leukocytes, which contained less than 1.5% of the ³H-labeled DNA. The erythrocyte chromatin was prepared and fractionated as before to obtain a low-melting 3% fraction A and the remaining 97% fraction B. Unlabeled chromatin was fractionated in a parallel experiment to obtain two similar fractions. DNA was extracted and purified from each fraction. In the case of the unlabeled DNA, ³H was introduced *in vitro* by photoreduction with [³H]NaBH₄. The four labeled DNA samples were then fragmented to a size appropriate for hybridization, 1.7×10^5 single-stranded molecular weight, by limited depurination (Grouse *et al.*, 1972). The relationship of the base sequence of these four DNA's to red cell RNA was then investigated by hybridization assays.

Hybridization mixtures contained low concentrations of labeled single-stranded DNA and high concentrations of RNA to permit DNA-RNA hybrid formation without excessive DNA renaturation. The data in Fig. 6 demonstrate that the DNA from the low-melting fraction reacts with RNA to about 30%, while the other major fraction exhibited little or no complementarity with red cell RNA.

This experiment does not bear on the question whether this sequence complementarity is specific for homologous red cell RNA. For this reason the same experiment was performed with chicken liver RNA. In this case,

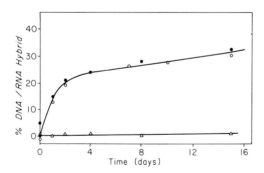

FIG. 6. Formation of DNA–RNA hybrids between DNA extracted from chick erythrocyte chromatin fractions and chick erythrocyte RNA. DNA, 4 mg, from *in vitro* ³H-labeled (●) or *in vivo* ³H-labeled (○) erythrocyte chromatin fraction A or *in vivo* ³H-labeled fraction B (△) was incubated with 8 mg of erythrocyte RNA. The incubations were carried out in 0.4 ml of 0.4 M phosphate buffer (pH 6.8) at 70°C. At various times, 50-μl aliquots were removed and diluted to 0.14 M phosphate buffer, and the amount of DNA-RNA hybrid was assayed by hydroxyapatite chromatography (Grouse *et al.*, 1972).

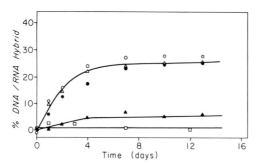

FIG. 7. Formation of DNA-RNA hybrids between chicken DNA or DNA extracted from chicken erythrocyte chromatin fractions and chicken liver RNA. DNA, 4 µg, from ³H-labeled fraction B (△) or ³H-labeled fraction A (▲) was incubated with 9 mg of liver RNA in 0.4 ml of 0.4 M phosphate buffer at 70°C. Incubations were similarly carried out with liver RNA and total DNA (○) or unique DNA sequences (●). The liver RNA *in vivo* ³H-labeled fraction B mixture was also treated with alkali to hydrolyze the RNA prior to incubation (□). The amount of DNA-RNA hybrid was assayed by hydroxyapatite chromatography.

the ³H-labeled DNA from low-melting red cell chromatin reacted to only about 4% with liver RNA. Conversely the DNA of the higher-melting fraction did react to an extent of more than 20% with liver RNA (Fig. 7).

Essentially the same saturation value was obtained for hybridization with liver RNA when total DNA or unique sequences purified from total DNA were used (Fig. 7). This is to be expected if most of the sequences transcribed in liver are located in the major 97% fraction of erythrocyte chromatin.

One further control experiment is necessitated by the very high ratio of RNA to DNA used in these experiments. With such a large amount of RNA, it is possible that the reaction is really due to contaminating unlabeled DNA. That this is not the case is deonstrated by the absence of any reaction when the liver RNA preparation was treated with alkali (Fig. 7).

VI. CONCLUSIONS

Several attempts to fractionate chromatin preparations have been made over the past few years. Methods based upon differential centrifugation of sheared chromatin have been useful for separating active and inactive segments of thymocyte chromatin (Frenster *et al.*, 1963; Frenster, 1969) and more recently for isolating fractions containing highly redundant DNA (Yunis and Yasmineh, 1971; Duerksen and

McCarthy, 1971). This same method has been successful for *Drosophila* chromatin, where the fraction of active chromatin is relatively large (McCarthy *et al.*, 1973). However, in our hands the resolution has proved limited for vertebrate chromatin.

Several other fractionation approaches have been successful, including enzymatic digestion (Billing and Bonner, 1972), chromatography on ECTHAM cellulose (Reeck *et al.*, 1972), and agarose gel filtration (Janowski *et al.*, 1972).

In the case of chick erythrocyte chromatin, the thermal fractionation of chromatin appears to be quite efficient. The DNA of the major fraction, 97% of the total gives little or no reaction with red cell RNA. On the other hand, the smaller fraction contains DNA which reacts to almost 30%. Assuming one-strand transcription, this implies that about 60% of this DNA is complementary to RNA. Alternatively stated, the active DNA has been enriched some 20-fold as a result of this simple procedure. This part of the genome, although complementary to erythrocyte RNA, reacts only poorly with liver RNA. Indeed the low reaction may result from the presence of red blood cells in the liver. Thus, one may conclude that the fractionation depends on the state of the genome in a particular cell rather than an intrinsic property of those DNA segments tending to make them low melting. This conclusion must be qualified slightly, however, since other experiments did indicate that active DNA in chicken red cells is of lower than average G + C (McConaughy and McCarthy, 1972). This could contribute to the low-melting behavior, although a G + C effect alone could not account for a T_m 15°C below that of total chromatin. Unfortunately, we are as yet unable to ascribe the lower T_m to any difference in chemical composition of active chromatin. In accordance with currently popular models of chromatin structure, it could be due to the presence of acidic proteins neutralizing histone binding and stabilization, the complete absence of histones or modification of histones by methylation, acetylation, or phosphorylation. Resolution of this issue must await the application of other fractionation methods.

The same thermal fractionation methodology has been applied to sea urchin embryo chromatin. In this case it was possible to show that pulse-labeled blastula RNA, containing a high proportion of histone mRNA, specifically hybridized with the DNA isolated from low-melting chromatin (M. M. Farquhar and B. J. McCarthy, unpublished results). Thus it appears that the method may have general applicability to the isolation of transcriptionally active DNA from any eukaryotic cell.

Segments of chromatin active in RNA synthesis exhibit a lower thermal denaturation temperature, and the DNA moiety of these segments may be purified by thermal elution. It is obvious that the method is inappropriate for characterization of chromosomal protein in chromatin fractions since high temperature and the hydroxyapatite itself leads to denaturation and poor recovery. Furthermore the high temperature and relatively high ionic strength would tend to promote some exchange of chromosomal protein (Clark and Felsenfeld, 1971).

VII. SUMMARY

The prospects of separating active and inactive segments of chromatin are discussed. In many vertebrate cells the problem appears to be difficult, since the fraction of active chromatin is small and chemical and physical properties may be quite similar to those of the inactive segments. Nevertheless several methods have proved to be successful, including thermal chromatography on hydroxyapatite discussed here in some detail.

ACKNOWLEDGMENTS

The research described above was supported by grants from the USPHS (4960) and the National Science Foundation (4851).

REFERENCES

BASERGA, R., and STEIN, G. (1971). Neoplasia and the cell cycle. *Fed. Proc., Fed. Amer. Soc. Exp. Biol.* **30**, 1752–1759.

BILLING, R. J., and BONNER, J. (1972). The structure of chromatin as revealed by deoxyribonuclease digestion studies. *Biochim. Biophys. Acta* **281**, 453–462.

BONNER, J., CHALKLEY, G. R., DAHMUS, U., FAMBOUGH, D., FUJIMURA, F., HUANG, R. C. C., HUBERMAN, J., JENSEN, R., MARUSHIGE, K., OHLENBUSCH, H., OLIVERA, B., and WIDHOLM, J. (1968). Isolation and characterization of chromosomal nucleoproteins. *In* "Methods in Enzymology" (L. Grossman and K. Moldave, eds.), Vol. 12, Part B, pp. 3–64. Academic Press, New York.

BRITTEN, R. J., and KOHNE, D. E. (1968). Repeated sequences in DNA. *Science* **161**, 529–540.

BROWN, I. R., and CHURCH, R. B. (1971). RNA transcription from non-repetitive DNA in the mouse. *Biochem. Biophys. Res. Commun.* **42**, 850–855.

BRUTLAG, D., CAMERON, S., and BONNER, J. (1969). Properties of formaldehyde-treated nucleohistone. *Biochemistry* **8**, 3214–3218.

CLARK, R. J., and FELSENFELD, G. (1971). Structure of chromatin. *Nature (London), New Biol.* **229**, 101–105.

DAVIDSON, E. H., and HOUGH, B. R. (1971). Genetic information in oocyte RNA. *J. Mol. Biol.* **56**, 491–506.

DUERKSEN, J. D., and McCARTHY, B. J. (1971). Distribution of DNA sequences in fractionated chromatin. *Biochemistry* **10**, 1471–1477.

FIRTEL, R. A. (1972). Changes in the expression of single copy DNA during development of the cellular slime mold *Dictyostelium discoideum*. *J. Mol. Biol.* **66**, 363–377.

FRENSTER, J. H. (1969). Biochemistry and molecular biophysics of heterochromatin and euchromatin. *In* "Handbook of Molecular Cytology" (A. Lima-de-Faria, ed.), pp. 251–276. North-Holland Publ., Amsterdam.

FRENSTER, J. H., ALLFREY, V. G., and MIRSKY, A. E. (1963). Repressed and active chromatin isolated from interphase lymphocytes. *Proc. Nat. Acad. Sci. U.S.* **50**, 1026–1032.

GELDERMAN, A. H., RAKE, A. V., and BRITTEN, R. J. (1971). Transcription of non-repeated DNA in neonatal and fetal mice. *Proc. Nat. Acad. Sci. U.S.* **68**, 172–176.

GROUSE, L., CHILTON, M. D., and MCCARTHY, B. J. (1972). Hybridization of RNA with unique sequences of mouse DNA. *Biochemistry* **11**, 798–805.

HAHN, W. E., and LAIRD, C. D. (1971). Transcription of non-repeated DNA in mouse brain. *Science* **173**, 158–160.

HUANG, R. C., BONNER, J., and MURRAY, K. (1964). Physical and biological properties of soluble nucleohistones. *J. Mol. Biol.* **8**, 54–64.

ITZHAKI, R. F., and COOPER, H. K. (1973). Similarity of chromatin from different tissues. *J. Mol. Biol.* **75**, 119–128.

JANOWSKI, M., NASSER, D. S., and MCCARTHY, B. J. (1972). Fractionation of mammalian chromatin. *In* "5th Karolinska Symposium on Research Methods in Reproductive Endocrinology" (E. Diczfalusy, ed.), pp. 112–129.

MCCARTHY, B. J., and CHURCH (1970). The specificity of molecular hybridization reactions. *Annu. Rev. Biochem.* **39**, 131–150.

MCCARTHY, B. J., NISHIURA, J. T., DOENECKE, D., NASSER, D. S., and JOHNSON, C. B. (1973). Transcription and chromatin structure. *Cold Spring Harbor Symp. Quant. Biol.* **38** (in press).

MCCONAUGHY, B. L., and MCCARTHY, B. J. (1972). Fractionation of chromatin by thermal chromatography. *Biochemistry* **11**, 998–1003.

MIYAZAMA, Y., and THOMAS, C. A., JR. (1965). Nucleotide composition of short segments of DNA molecules. *J. Mol. Biol.* **11**, 223–237.

OLIVERA, B. M., BAINE, P., and DAVIDSON, N. (1964). Electrophoresis of the nucleic acids. *Biopolymers* **2**, 245–257.

REECK, G. R., SIMPSON, R. T., and SOBER, H. A. (1972). Resolution of a spectrum of nucleoprotein species in sonicated chromatin. *Proc. Nat. Acad. Sci. U.S.* **69**, 2317–2321.

TURNER, S. H., and LAIRD, C. D. (1973). Diversity of RNA transcription sequences in *Drosophila melanogaster*. *Biochem. Genet.* **10**, 263–274.

YUNIS, J. J., and YASMINEH, W. G. (1971). Heterochromatin, satellite DNA and cell function. *Science* **174**, 1200–1209.

Gene Manipulation: Progress and Prospects

Ethan Signer

Department of Biology, Massachusetts Institute of Technology,
Cambridge, Massachusetts

I. INTRODUCTION

The recent explosive development of the biological sciences will soon allow humanity to manipulate its own genetic material. Our manipulation of the external environment in the past probably indicates both our readiness to do so and the abuses that may be expected. At present only the genes of microorganisms may be manipulated easily, but the rapid progress in this area makes it essential to begin considering the implications for the manipulation of human genes.

I shall summarize briefly the progress with microbial systems; discuss some likely applications to humans; explore a current example; outline the nature of the problem posed by gene manipulation; and propose steps toward a solution.

II. PROGRESS WITH MICROBIAL SYSTEMS

Several techniques are available for moving bacterial genes around. In a few species, naked DNA may be introduced into a cell from outside and permanently incorporated into the chromosome (for general bacterial genetics, see Hayes, 1968; Beckwith and Zipser, 1970). In many species DNA can instead be transferred from one bacterium to another by being wrapped up in the coat of a bacterial virus instead of the virus' own

DNA. It may then be incorporated stably into the chromosome of the second bacterium ("general transduction").

The DNA of some bacterial viruses may itself become inserted into the host chromosome. In certain conditions bacterial genes may become permanently attached to the DNA of such a virus, making what is essentially a new, genetically stable virus carrying both viral genes and a few bacterial genes. The DNA of this new virus may then be inserted more or less permanently into the chromosome of a second bacterium ("special transduction"). Similar manipulations can be done with, instead of viruses, other DNA elements (episomes) that exist either in the cell cytoplasm or added to its chromosome, and may be transferred to other cells by contact, with these techniques it is possible in principle to pick up essentially any gene of the bacterium *Escherichia coli,* incorporate it into the DNA of a virus or an episome, and then move it to any other location on the bacterial chromosome (Beckwith *et al.,* 1966; Gottesman and Beckwith, 1969).

The DNA of a gene (part of the lactose operon of *E. coli*) has been isolated in pure form and seen in the electron microscope, by techniques applicable in principle to any bacterial gene (Shapiro *et al.,* 1969). Similarly, a gene (for yeast alanine transfer RNA) has been chemically synthesized, by methods valid in principle for any gene of known sequence (Agarwal *et al.,* 1970). Techniques like these might ultimately allow specifically modifying genes in the laboratory, and then introducing them back into the organism.

Besides the structure of genes, their function may also be manipulated. Mutating regulatory proteins, such as repressors, or the sites of DNA (operators) where they act, may affect when a gene is or is not transcribed into messenger RNA; and mutating the sites on DNA (promoters) where messenger RNA is initiated may change the maximal rate of messenger synthesis. Different subunits of the enzyme RNA polymerase may affect the way the enzyme chooses which genes to transcribe into messenger, and small molecules, such as antibiotics, may affect both transcription of genes into messenger RNA and translation of messenger RNA into protein.

III. POSSIBLE APPLICATIONS TO HUMANS

Model microbial systems raise a number of obvious and not so obvious possibilities for manipulating human genes.

Some technical research concerns how to manipulate cultured mammalian cells as easily as bacteria. One report, apparently not yet verified, claims the introduction of at least one gene into cultured human cells as naked DNA and its incorporation into the cell's chromosomes (Szybalska and Szybalski, 1962). It might also be possible to insert genes from one species into the chromosome of another. When chick blood cells are fused with mouse fibroblast cells, the resulting hybrid cell seems to retain only the mouse chromosomes, yet to have at least one chick gene, presumably incorporated into a mouse chromosome (Schwartz *et al.*, 1971). Similarly, chicken cells are reported stably to incorporate mouse DNA from the outside (Hill and Hillova, 1971), and barley root cells stably to incorporate bacterial DNA (Ledoux and Huart, 1969).

Like some bacterial viruses, polyoma virus particles occasionally carry host (mouse) DNA rather than polyoma DNA. It is not yet known whether this DNA can become incorporated into the genome of a second mouse cell infected by these particles, as in bacteria (general transduction) (Osterman *et al.*, 1970). Mammalian viruses such as SV40 (and probably polyoma) can insert their DNA into the chromosomes of the host cells (Sambrook *et al.*, 1968). Since purified SV40 DNA is infective, an isolated or synthetic human gene might be attached to it chemically so that after infection the gene is inserted into the host chromosomes together with the viral DNA (special transduction). Alternatively, the isolated messenger RNA for a particular gene might be added chemically to all or part of the RNA of a tumor virus, known to be taken in from the outside by mammalian cells and then transcribed into a DNA copy that is added to the genome of the host cell (Temin and Mitzutani, 1970; Baltimore, 1970).

Rather than altering the genes themselves, it might also be possible to turn genes on and off highly selectively, using hormones (Tomkins *et al.*, 1969).

Other technical research concerns how to intervene in prenatal human embryonic development. Amniocentesis—removal of some of the amniotic fluid from the uterus of pregnant women so that the cells of the developing embryo may be examined—is already being used clinically in counseling parents about prospective genetic disorders (Milunsky *et al.*, 1970). Artificial insemination of females is quite common in cattle breeding and is also used for humans (Jones, 1971). Since sex in mammals is determined only by whether the sperm carries an X or a Y chromosome, separating X-bearing from Y-bearing sperm will soon allow choosing in advance whether insemination will produce a male or a female.

The female might be dispensed with entirely, except as a source of eggs. The eggs might be fertilized in the laboratory, and the resulting embryo either allowed to develop entirely in an artificial uterus or reimplanted in the uterus of a female. Human eggs have been fertilized in the laboratory and the embryos cultivated for some time (Edwards *et al.*, 1970). Mouse embryos have been fertilized *in vitro*, cultivated and successfully reimplanted in the uterus of a live animal (Whittingham, 1968), and the sex of rabbit embryos has been successfully determined before reimplantation (Gardner and Edwards, 1968).

Even fertilization of the egg will probably not be necessary in the near future. When the haploid nucleus of a frog egg is destroyed or removed, and replaced with a diploid, mature nucleus from the intestinal cell of a tadpole, the egg can then develop like a normal embryo into a normal-seeming adult that is an exact replica of the individual donating the nucleus (Gurdon and Uehlinger, 1966). Besides showing that the intestinal nucleus carries all the information for generating an entire individual, this means that a single adult can give rise to many identical "twins" without needing sex. This way of asexually reproducing animals, called *cloning*, is now being extended to mammals. Once replacement of the haploid egg nucleus with a diploid mature nucleus—right now the limiting step for mammals—is mastered, cultivating the developing embryo as above will generate a mature individual by completely asexual means. A mammal will probably be cloned within the next ten years.

Embryos may be manipulated in other ways too. The cells of two or more early mouse embryos may be gently dissociated, and mixed together to reconstitute a single embryo that is a mosaic of the different cell types that make it up. Ultimately this embryo gives an adult mosaic individual (Mintz, 1967).

Some techniques have evident therapeutic applications. Diseases resulting from a failure to turn a gene on or off could be treated by hormonal or other ways of controlling gene expression. Hormone therapy is already used to treat diabetics, deficient in the hormone insulin.

Rather than giving diabetics insulin, one might supply them with genes leading to the synthesis of insulin (Sinsheimer, 1969)—either as purified DNA, or wrapped in a virus particle, or even as part of a virus that would attach to their chromosomes without harmful effects. Not only would the diseased individuals be spared daily insulin injections, but supplying their germ cells with the normal gene could remove diabetes from the population.

Many new possibilities will be opened up if DNA from other species can be stably incorporated into human cells. Genetic diseases could be treated with genes from bacteria, for instance, where genes and the proteins they produce may be characterized relatively easily, and where genes might be tailored for specific purposes by mutation or other manipulations.

Some of a virus' own genes might be used for therapy. The Shope papilloma virus is not known to produce clinical symptoms in humans, but it does produce an enzyme that destroys the amino acid arginine. The virus could be used to treat patients with a high level of blood arginine, said to be associated with mental retardation and other clinical abnormalities (Rogers, 1970).

Mixing the dissociated cells from a defective and a normal embryo might produce a mosaic embryo that would allow an otherwise inviable organism to survive. Perhaps one could mix in a few suitably modified cells at just the right stage of embryonic development so that the effects are quite specific. Adults might also be made partial mosaics—the blood-forming cells of certain anemics, for instance, might be removed, genetically tailored to compensate for the defect, and then replaced.

Fertilization of eggs in the laboratory and reimplantation in the uterus have obvious applications in curing infertility.

Besides therapy, gene manipulation has eugenic potential. We already practice eugenics on a very tiny scale with amniocentesis and genetic counseling. Control over the human population will be greatly increased when lines of sperm and egg can be selected for *in vitro* fertilization, and cultivation of embryos *in vitro* will allow better quality control. Cloning will allow society to have many identical copies of particularly desirable individuals or types. Early embryos might be artificially stimulated to subdivide and produce several identical embryos; one embryo might then be cultivated to term, while the others are kept in storage banks so that valuable people could be replaced in case of accidental death (although environmental factors would help determine the resulting adult). And, at a very modest level, selection of sperm will allow couples to choose to have a boy or a girl.

Besides its therapeutic and eugenic applications, gene manipulation makes possible new weapons. Research under way in America and presumably other countries as well already makes use of advances in bacterial genetics to construct militarily desirable strains of microorganisms, and much more sophisticated weapons of this kind might already be under development. Almost any virologist, for example, will acknowledge the feasibility in principle of isolating a "host range" mutant of a virulent virus of humans that will grow better in skin cells having more black or yellow pigment than those of "whites"—perhaps the ultimate in racism. And the more we learn about manipulating genes, the greater the possibility of designing particularly virulent or particularly selective strains.

But there are also less obvious and more insidious implications of gene manipulation. Mostly these stem from a basic problem: Who decides how and to whom gene manipulation is to be applied?

To choose a simple example, characterizing the chromosomes of an embryo by amniocentesis is used as a basis for counseling the elimination of fetuses with a genetic defect. But who decides what is a genetic defect? Most people might agree that a fetus with an extra chromosome-21 (Down's syndrome or mongolism) would be better off aborted, but other cases are not so clear. Males with two Y chromosomes (XYY) are said to be overly aggressive and have criminal tendencies. Aside from the question of penetrance—how many XYY's in the population would actually show an "antisocial" phenotype—perhaps the behavioral phenotype of an XYY is antisocial only in response to the particular stresses of our rather far from ideal society. And perhaps the XYY genotype also has some beneficial effect, just as the gene that causes sickle cell anemia also leads to increased resistance to malaria (Allison, 1954).

Control over fertilization raises more serious problems. If artificial fertilization is carried out on a large scale, who will select the lines of sperm or egg to be used? And with the sexing of sperm, even when couples use their own germ lines the effect on society would be drastic if, as is likely at present, most people chose to have boys rather than girls.

The technique of cloning would allow making people with particularly desirable characteristics. The government might clone itself an army of ideal soldiers; or an industrial corporation, a population of drones to work in the factories—or the academic community, a population of philosopher kings. Lines of people might be bred genetically pure for a few characteristics, so that inducing early embryos to subdivide into multiple copies might be an easier way of doing this. Cloning might, of course, be restricted to a few highly valuable individuals, but, say, the black ghetto and the present government might have rather different preferences as to whom to clone. Furthermore, traffic in, say, bootleg Einsteins for families trying to improve their social status might be profitable.

Concern about problems of this sort is far from premature. Verging on feasibility is Shockley's suggestion for birth control (cited in Fleming, 1969): all women are made reversibly sterile by implantation of a contraceptive beneath the skin, removable by a doctor only on presentation of a government license to have a child. One can imagine what might influence our government to grant or refuse a license.

Nor are subtle uses of science for social control only in the future. IQ tests of doubtful validity have been used for some time to help track children of poorer families from elementary school ultimately into the less desirable jobs. This application of genetics is not a logically inexorable extension of scientific research, but rather the somewhat forced use of one of the few areas of science even remotely relevant to the ulterior purpose—namely, the maintenance of a stable class structure in the interests of the dominant but not the weaker sectors of society. As we shall see, this way of using science is by no means uncommon.

Already medicine routinely decides how and to whom medical advances are to be applied (Ehrenreich and Ehrenreich, 1971). High medical technology—in fact, hospital care in general—is concentrated among the wealthier and middle classes, not the poorer classes, in whose neighborhoods the hospitals are often located.

In the same vein is the gradual but increasing development of the weaker sectors as a biological reservoir for the rest of society. The weaker sectors already serve as guinea pigs for medical research (for a highly establishment-oriented view, see *Daedalus*, 1969). Drugs and other remedies are frequently tried out, especially for potential side effects, on poor people, prisoners, and conscientious objectors—so-called volunteers who are usually under substantial pressure, monetary or otherwise, to comply—or else tested abroad in countries with even lower testing restrictions than ours. Birth control pills were first tested on poor Puerto Rican women, not on the middle and upper class women who would be the ones able to afford them when marketed. And it is common for technical innovations in medicine to be tried out on poor patients in hospital emergency rooms.

The weaker sectors are also coming to be an organ bank for transplantations (Zimmerman *et al.*, 1971). This is already true for blood, which is donated (often sold) primarily by the poorer sectors but used primarily by the wealthier sectors in this country (although not in Britain; see Titmuss, 1971). Probably the rights only to the organs of the recently dead will be sought at first, as with heart transplants at present. But it is not too much to imagine soliciting living organs from poor people—a kidney say, or a lung—if the price is right, since money is one of the few rewards the weaker sectors can aspire to in our society.

The trend will probably continue with the weaker sectors acting as guinea pigs for gene manipulation and prenatal intervention. Taking some rather extreme examples, if science ever attempts to make a child whose head is twice as big as normal simply for the sake of trying, as Crick speculates (*Time Magazine*, 1971), or a legless astronaut who would take up less room in a space capsule, as Haldane suggests (*Time Magazine*, 1971), the children will probably come from the weaker sectors. Once reimplantation of embryos is developed (after experiments on poor women, no doubt) it will cure infertility, but it will also permit wealthy women to hire poor surrogate mothers to go through their pregnancies, technologically updating the wet nurse.

Thus the therapeutic, eugenic, and military applications of gene manipulation overall will probably exacerbate social class distinctions and increase the possibilities for exploitation of the weaker sectors by the more powerful ones. These applications, hardly an exhaustive list, are paralleled in other areas of biology, notably brain research, with its increasing possibilities of social control through drugs and electronic stimulation (see, for example, *Time Magazine,* 1971). If these prospects seem fanciful, we might recall that in 1940 both the hydrogen bomb and the man on the moon were generally considered to be inconceivable within our lifetimes. And the applications cited here are by no means at the limit of imagination, as the science-fiction literature shows (e.g., Brunner, 1969).

IV. A CURRENT EXAMPLE

It is important to realize that science may be used by the dominant sectors of society against the interests of the weaker ones, not only intentionally, but simply as a by-product of technological development. An example of gene manipulation in this context is the teratogenic effect on humans of chemical spraying by the United States government in Vietnam.

The United States has sprayed Vietnam with chemicals since 1961, officially to defoliate plants. Some 22,000 km^2 are estimated to have been sprayed, of which about 11% is crop land and the rest forest, mostly in South Vietnam. Spraying kills food crops and trees (mangrove swamps are particularly sensitive, and 20% of the economically valuable hardwood of South Vietnam is estimated to have been destroyed); its effects lead to soil erosion and laterization, and severe changes, possibly irreversible, in animal and fish ecology (*Science,* 1971).

One of the chemicals, 2,4,5,-trichlorophenoxyacetic acid ("245T", used in combination with picloram as "Agent Orange") is teratogenic in mice and rats partly owing to the impurity dioxin, formed from 245T at high pH and temperature, although it is not yet clear whether 245T free of dioxin is also teratogenic (Courtney *et al.,* 1970; *Nature,* 1971). It has been estimated that a dose of Agent Orange toxic for humans could be ingested after heavy spraying and moderate rainfall in Vietnam by drinking 2 or 3 quarts of water (A. W. Galston, personal communication). 245T is now said to be banned by the US government from use in populated areas in both the US and Vietnam.

A high rate of stillbirths in Tay Ninh Provincial Hospital, and a rise in the incidence of the birth defects cleft palate and spina bifida in Saigon, have been reported coincident with large scale spraying of South Vietnam (*Science,* 1971). Physiological effects—nausea, vomiting, headaches, weakness lasting up to several months—are said to be widespread after spraying. A study of 903 South Vietnamese patients who had been evacuated to the Benh Vien Hospital in Hanoi (Tung *et al.,* 1970) included 179 patients who had lived from 2 months to 5 years in sprayed areas or had themselves been directly sprayed. Of these, 19 were women, 4 of whom had been pregnant when sprayed, or living in sprayed areas. One of the 4 children was born normal, the other 3 abnormal. The mothers appeared normal (aside from the effects of spraying) and had received neither medication nor X-rays during pregnancy, and congenital abnormalities were said to be absent in the three families over at least the past three generations.

The first child, a 3-year-old girl, was born to a 37-year-old woman who had lived in a defoliated zone for 4 years. The child has the symptoms (Figs. 1 and 2) and karyotype (Fig. 5) of Down's syndrome (mongolism).

The 23-year-old mother of the second child lived for 2 months in a defoliated zone and was sprayed at 7 weeks of pregnancy. The child, a 10-month-old girl (Fig. 3) is a microcephalic, has an abnormality of the palate, and suffers from convulsions.

The 38-year-old mother of the third child previously had two normal children. She was sprayed directly twice before pregnancy and a third time during the 7th week of pregnancy, all while working in the rice paddies. The child (Fig. 4), a 17-month-old girl, has an abnormally shaped head and extremities. Her left foot (Fig. 6) has 6 toes, 3 abnormally long, and her thumbs are spatulate—significant because digital differentiation occurs in human embryos at about the 7th week of pregnancy, precisely the time when her mother was sprayed. The child is mentally retarded. When I saw her at age 23 months she could neither walk nor talk.

These data were not presented by the Vietnamese as conclusive (particularly considering the age of two of the mothers) but rather as highly suggestive. The important fact is not the certainty, but simply the strong possibility, of teratogenic effects from chemical spraying.

It is quite possible that the US government would not have used 245T if it had been known to be teratogenic, not necessarily because of humanitarian considerations but because the publicity is unfavorable. (On the other hand, a North Vietnamese scientist asked me if the US government is likely to increase the concentration of dioxin now that it is a confirmed teratogen.) Also, the example was not selected to argue for a cleaner war with conventional weapons.

The point is that regardless of intent, this kind of tragedy is inevitable. If science is used primarily in the interests of the dominant sectors, frequently in direct conflict with the interests of weaker sectors or countries, we should not really be surprised when that use occasionally exceeds the limits of good taste. All the details are typical—the war against the Indo-

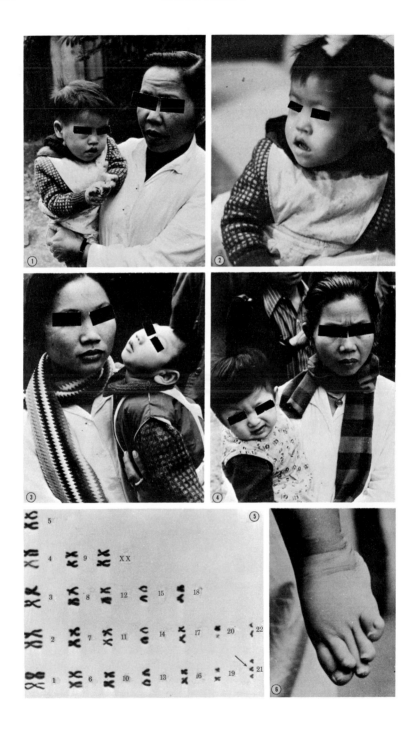

chinese peasants in the first place; the unprecedented use of technological and scientific sophistication in weaponry; the government's reluctance to test the defoliant chemicals when their teratogenic effects began to be suspected from results in the field, and later to release the test findings; the mildness of the response of the American public and press; and the continued, unabated use of other defoliants and sophisticated technological weapons. Other examples abound. If not teratogenesis in Vietnam, then thalidomide babies (and it is well to recall the struggle that was required to achieve acknowledgment of the problem and to remove thalidomide from the market), and nuclear fission in Hiroshima and Nagasaki, and health hazards in American industry, and environmental pollution, and other similar scientific fallout from technological society.

As things now stand, larger-scale gene manipulation may simply be next on the list.

V. NATURE OF THE PROBLEM POSED BY GENE MANIPULATION

We can foresee, then, some rather serious and undesirable applications resulting from advances in gene manipulation. How can these—and, for that matter, any undesirable application of science—be avoided?

Clearly, scientists can avoid doing research leading to harmful application, but this is barely a minimal solution. Arthur Galston's Ph.D. thesis research showed that 2,3,5-triiodobenzoic acid (TIBA) increases the number of soybean flowers formed, leading to more pods and a bigger harvest. Although used commercially to increase crop yields, TIBA at somewhat higher concentrations was later shown also to cause the buds and leaves of plants to fall off, a finding that helped government chemical warfare researchers at Fort Detrick design defoliating chemicals for Vietnam. Galston's contribution, then, unwittingly and very much against his will, was transformed from alleviating hunger to destroying vegetation in a war he considers to be immoral (Galston, 1971). Thus the application of one piece of research may be both good and bad, and in any case unpredictable.

Figs. 1 and 2.   Three-year old girl with Down's Syndrome, whose mother had lived in a defoliated zone for four years.

Fig. 3.   Ten-month old microcephalic girl, mother sprayed with defoliants during seventh week of pregnancy.

Fig. 4.   Seventeen-month old girl with abnormalities of head and extremities, mother sprayed before and during seventh week of pregnancy.

Fig. 5.   Karyotype of child shown in Figs. 1 and 2.

Fig. 6.   Left foot of child shown in Fig. 4; note length and number of toes.

Traditionally, scientists resolve this sort of problem with an implicit (and sometimes explicit) faith that responsible forces in society will not allow science to be applied for evil. Nazi Germany, where the Good Germans looked the other way while Jews, gypsies, and other "misfits" in Hitler's (*genetic*) scheme were taken to the gas chambers, would seem to show that under the proper conditions society will allow anything. Yet we can still find Francis Crick saying (*Time Magazine*, 1971), of the possible evils of gene manipulation, "people will simply not stand for them."

By Crick's argument, one would have to say that black and brown people are "standing for" conditions in the ghetto; miners in Appalachia are "standing for" black lung, unemployment and loss of self-respect; American Indians are "standing for" extermination of their culture; underdeveloped nations are "standing for" the plunder of their resources; and the Indochinese peasants are "standing for" the destruction of their nations and societies, to cite only a few cases. At issue is the question of social control: Who determines how society develops, and therefore what the predominant character of society will be? These examples—and there are many more—suggest that many sectors of society do not have enough control over their own status, enough of a say in the direction society develops, to be able to determine for themselves what they will or won't "stand for," including applications of science harmful to them.

Therefore, we cannot really address the prospective effects of gene manipulation without first considering the role of science in society.

I shall start from several assumptions that have been documented extensively (Mills, 1956; Domhoff, 1967; Lundberg, 1968; Baran and Sweezy, 1966; Horowitz, 1965; Magdoff, 1969). They are that the dominant ethics in American society—particularly, that it is legitimate for one person to profit directly from another person's labor, and that the means of production for society's needs may be privately owned—result inevitably in a severely disproportionate concentration of wealth and power in the hands of a tiny minority of society; that this minority and its supporting sectors invariably exploit both the many weaker sectors in America and also the weaker foreign countries, and that a large part of the exercise of power by the dominant sectors is concerned with maintaining their position of strength.

I shall argue, then, that science can be viewed in the following way. *Power is exercised by the dominant minority principally in the ever-expanding development of technology. The main result is to widen the gap between the managerial and the weaker sectors, so that technology functions more and more in the interests of the managerial and only incidentally in the interests of the weaker sectors. Science functions primarily to support technology. The general direction of scientific research is consistent with the needs of technology, although within these limits there is considerable flexibility and permissiveness for individual endeavor. The willing, although not necessarily informed, participation of scientists in this arrangement is purchased with rewards that, compared with those available to most of society, are munificent.*

Thus it makes no difference whether or not physicists intend or want their basic research to be used for weapons and surveillance devices; mathematicians, geophysicists, and metallurgists for ballistic missiles; biologists, chemists, and medical researchers for bacteriological weapons, defoliants, and "riot control" gases; and anthropologists and sociologists for counterinsurgency research, or whether these and other scientists are content to have science used for social control (Zimmerman *et al.*, 1972). *The way science is used is determined by the interests and orientation of the dominant social sectors, and there is no reason to expect this to be any less true for gene manipulation than other research.* Therapeutic applications, for instance, will probably be developed, but according to the present pattern such benefits will be controlled by and available mainly to the dominant sectors, and will help alienate them from the weaker sectors. And it is virtually certain that any applications with a potential for exploitation, weaponry, or any other means of social control will also be developed, and most probably used, for just that purpose.

To support these assertions I shall consider briefly the roles in society of technology, of science, and of the scientist, and the control of scientific research (see also Zimmerman *et al.*, 1971; Beckwith, 1970).

Historically, science after the Dark Ages was primarily a search for truth and an attempt to comprehend the universe, and as part of the developing rationalist ethic it was an intellectually progressive and liberating force. The transformation of science from a detached intellectual pursuit to a commodity that can be used to maintain and extend power was made possible by the rise of technology.

Technology has been dealt with extensively by McDermott (1969), whose argument may be summarized as follows. The usual definition of technology as the organization of knowledge for practical purposes does not specify who does the organizing. Technology is more properly defined at present as an institutional system for highly centralized and intensive control over large groups of men, machines, and events by small groups of technically skilled men operating through organizational hierarchies. Technological organization defines the roles and values of its members, not vice versa. Therefore, as technology has developed it has progressively widened the gap—in culture, quality of life, and power to control society—between the weaker sectors and the dominant managerial ones.

Rather than serving society, then, technological organization progressively minimizes the participation of the weaker sectors in society.

Technology, he continues, has combined centralization with mystification. Although the problems raised by technology are essentially social ones, the real causes usually become obscured as technocrats try to solve them by applying more technology—for example, in the cities and ghettos, where technology is not the cause of the problem but rather the agent of the real cause, namely, social and economic exploitation. And negative aspects of technology are increasingly underplayed as the survival of technological institutions comes to depend more and more on the positive aspects.

Science and technology are closely related with considerable overlap of institutions and personnel, and much of the foregoing applies directly to science too. Clearly science provides the fundamental knowledge on which technological advances depend. Similarly, science provides the body of knowledge and personnel essential for training the technicians. But beyond these basic necessities, science serves technology in several other ways.

By their close relation, science as an intellectual endeavor probably tends to raise the intellectual level of technology, thereby increasing somewhat the rate of its development. Similarly, science tends to endow technology with some of the mystery and impenetrability with which it cloaks itself. The public has always viewed science with some fear and awe, as magic that ordinary people can never be privy to, and this probably helps protect science and technology somewhat from too close public scrutiny and control.

Although science is often done without considering application, once a basic discovery is to hand its application as technology is almost inevitable (discussed below). Having a separate category of science helps technology by serving as the locus of certain kinds of basic research—useful for weapons, for instance, or perhaps gene manipulation—that might be socially unacceptable done as mission-oriented technological research and development. This arrangement nicely suits the interests of the scientists themselves, who in most cases explicitly disclaim responsibility for applications of their work.

Science also helps support some of the fundamental ideological tenets of technological society, namely that the benefits of society, including science, are distributed equitably among all sectors; that science is a strictly rational, objective search for truth; that scientific truth is inherently beneficial; that science is neutral and value-free; that scientists are more objective and less self-seeking than other people. These would seem to be, at best, gross oversimplifications, functioning primarily as myths that rationalize the real situation rather than accurate descriptions of it.

Finally, science serves technology by providing an accepted avenue to power for individual scientists. Having surmounted the obstacles of ever more selective examinations stretching from kindergarten through the Ph.D., scientists are finally admitted to the exclusive elite permitted to find (or at least seek) self-fulfilment doing exactly what gives them pleasure, with society's overt support. Thus scientists form one of the few sectors whose members control the planning, organization, and everyday details of their own creative work, rationally the prerogative of all sectors of society. Intellectual challenge is almost limitless; professionalism comfortably insulates scientists from intellectual, cultural, and physical contact with most other sectors; and a glamorous image is maintained (e.g., Fleming, 1969, 1970; Wilson, 1970; or News and Views in almost any issue of *Nature* since 1969) even though in reality science can be no less petty and self-serving than any other dominant area of technological society (Watson, 1968).

Power and prestige both within and outside science are also available. Competition for recognition by fellow scientists is keen, and scientific discoveries, at least of the right kind, are reverently publicized in the *New York Times* and elsewhere outside the field. The entrepreneurial image and skills are honed by the funding system, which encourages scientists to conceptualize research in commercial terms: competing for scarce resources; striving to reduce the time from raw materials to the finished published product; advertising themselves and their work; and viewing career advancement as a successful return on investment. Together recognition and entrepreneurship provide clear channels, often by way of research empires, to the higher echelons of industry and even government for scientists with a modicum of intelligence, ingenuity, and industriousness (see, for example, Greenberg, 1967).

With so many rewards, there is correspondingly little pressure on scientists to challenge the current image and values of science. To the contrary, concern for social responsibility is rightly seen by most scientists as a threat both to the largesse of funding and prestige dispensed by technocracy and to the pleasure of carefree intellectual endeavor (see Rose, 1970), which make the scientific mandarinate such a comfortable and satisfying sector of society.

All this is not to deny the cultural role of science as an intellectual pursuit, nor the value of a basic understanding of the universe. But science probably affects society much more through support of a technology developing according to its own interests.

How, then, is technology involved in controlling the direction of scientific research? Research depends largely on priorities—which problems are defined to be important, how and by whom, although it may also lead in unexpected directions. At least three factors keep the general trend of science in conformity with technological development.

The first is funding. In industry, of course, the power elite directly controls the funding of research. Through Congressional appropriations it also controls government research priorities fairly directly, but those of university research only in a very general way. [However, universities account for only about 10% of all research and development, about half of which is basic research, which in turn represents about half of all basic research (Lewontin, 1970)].

Influential scientists constitute a second factor. As they rise in the hierarchy of the scientific establishment, scientists exert increasing control over research both through panels that review funding of research grants and through their personal influence in defining for the scientific community which problems are important. But in turn their values and interests become more and more closely identified with those of the dominant sectors through their increasing social and professional rewards, distributed ultimately according to the values of the dominant minority.

The third factor is feasibility. Research will be done if it is feasible to do it, almost regardless of any other considerations. Edward Teller (1971), for example, says, "I believed in the possibility of developing the thermonuclear bomb. My scientific duty demanded exploration of that possibility." Importance and interest strongly influence which feasible research is supported. Importance is defined according to the accepted technocratic values, often mainly utility in overall technological development. Interest is very largely defined by intellectual challenge, also by no means independent of technocratic values—medical research, for instance, defines solving cancer, stroke, and heart disease, found mainly in the dominant sectors, as inherently more challenging than solving parasitic diseases, trachoma or malnutrition, diseases of most of the world's population, who tend not to live long enough to suffer cancer or stroke. However, both intellectual challenge and the reward of peer recognition for difficult achievements strongly motivate individual scientists to undertake difficult work especially, thereby tending to expand the limits of feasibility, often in unpredictable directions. Therefore, while not investing unprofitably, technology must support some research at the frontier of feasibility and stand ready to capitalize on scientific breakthroughs, just as it capitalizes on unexpectedly important research done only because it was feasible. Biology is an excellent example. Minimally emphasized and supported during the first half of the century, since the breakthrough of the 1950's and early 1960's the field has been rapidly integrated into the technocratic structure, with research expanding and feeding into the health and agriculture industries, and biologists attaining prominence in science, industry, and sometimes government. And on a small scale, the flexibility of technology generally seems to be increasing, judging from the shorter and shorter time lag from the laboratory to the marketplace and the battlefield.

It is not hard to find examples of science and technology functioning primarily in the interests of the stronger sectors with only incidental benefits to the weaker sectors, even excluding such extravaganzas as the man on the moon and the Indochina War. The Massachusetts Institute of Technology, one of the pillars of technocracy, has a School of Industrial Management, but no one is surprised at the absence of a School of Industrial Labor. The social scientific establishment studies how poor and ghetto populations function, and how to channel "disadvantaged" children through the educational system; but society provides no facilities for poor people to research how the powerful function, or how to educate their children to beat the channeling system and eventually obtain the desirable, well-paying jobs. Health care in America is curative rather than preventive, and strongly biased toward those who can afford it. The medical profession concentrates more and more on the dominant sectors at the expense of the weaker ones; medical technology becomes more and more sophisticated and expensive; and the excesses of the drug industry are widely discussed (Ehrenreich and Ehrenreich, 1971; Michaelson, 1971). Bluntly put, the dominant sectors have their science and technology, but the weaker sectors do not have theirs. And at the international level, the 6% of the world's population represented by the United States uses over half the world's yearly consumable resources. It is safe to say that, if the remaining 94% of the world's population—many of them hungry, diseased, and illiterate—disposed of comparable science and technology to ours, the situation would be rather different.

Projections for gene manipulation conform to this general picture, and it is in this context that we must consider the problem.

VI. STEPS TOWARD A SOLUTION

In the light of this analysis, avoiding the undesirable applications of gene manipulation or any other scientific research is only part of a larger problem: What must be done so that science benefits, not primarily the dominant sectors at the expense of the weaker ones, but rather all sectors of society? Gene manipulation is a useful paradigm because it might very soon have important social effects, but any solution to the problem of its application will have to embrace science as a whole.

Naturally scientists, like anyone else, should refuse work that has harmful consequences. But Galston's example shows how limited this approach really is, and in any case the research will often be taken up by others. Refusals of this sort, however, often focus on problems and bring them to the attention of the community. Two recent examples are Richard Lewontin's resignation from the National Academy of Sciences because of its unwillingness to stop sponsoring secret government research, and James Shapiro's decision, much publicized (e.g., Glassman, 1970) after his participation in isolating the lactose gene, to quit basic genetic research because of the potentially harmful consequences of gene manipulation.

With sporadic exceptions, the response of the scientific community has not been encouraging. Concerns about social responsibility are most commonly construed as attacks on the profession, provoking visceral closing of the ranks in defense and panic lest the golden Congressional funding goose be frightened to death. The scientific establishment—particularly the journal *Nature*—reacted to Shapiro's "defection" mainly in this tone, and seldom treated seriously the issues he raised rather than the nature of his action (e.g., *Nature*, 1970a,b; *Science*, 1970a,b).

Another common response is to acknowledge the problem, but then depend on the good will of scientists to solve it. Scientists "best realize the serious nature of the problems that beset us, . . . might be expected to lead the way in finding solution, . . . and . . . (are) most likely to rise above the limitations of national and ethnic prejudice and speak in the name of mankind as a whole" (Asimov, 1971). The role of scientists in technological society precludes taking this sort of proposal seriously, and the last clause in particular is doubtful considering that scientists are no less racists, sexist, and jingoist than any other group.

According to many liberal critics, ". . . where science and technology have expanded the problems, it may be only more scientific understanding and better technology that can carry us past them" (Platt, 1969; cf. Weinberg, 1970). Earlier I noted that the problems although raised by technology are fundamentally social, and not technical. No amount of our presently structured science and technology is likely to solve them, and more will just aggravate them. The fallacy of this position is coming to be recognized, although usually not with a very clear analysis (Siekevitz, 1970; Luria, 1972; see Rose, 1970).

In the particular case of gene manipulation, the scientific establishment usually ignores social effects and pointedly restricts discussion to technical questions (e.g., *Nature, New Biology*, 1971a,b). In the few instances where it is acknowledged that there might be effects on society, these are usually heavily underplayed (e.g., Davis, 1970; Lederberg, 1970).

Recently the National Institutes of Health sponsored a Conference on the Prospects of Gene Therapy. (This term has been proposed to replace "Genetic Engineering" which "has the revolting connotation . . . of impersonal scientific manipulation . . . and offends the dignity of many" (Aposhian, 1971); changing the name seems easier than dealing with the problem). All the 17 papers presented were to deal solely with technical aspects, and few of the participants felt impelled to raise any questions about social consequences. Of course, the NIH will soon hold a second meeting on social implications, but *not* for the scientists who participated in the first one. We should not be surprised at yet another instance of the technocracy sending the scientists off to play while taking care to minimize any assumption of social responsibility.

Even the occasional serious attempt to bring the problem into the open is not treated seriously by the scientific establishment. The firsthand reports of the recent Conference on the Social Impact of Modern Biology of the British Society for Social Responsibility in Science, by the usual anonymous correspondent in the establishment journal *Nature* (1970c) and by Beckwith (1971) in the radical journal *Science for the People*, might almost describe two different conferences. Both agree, however, that most of the established scientists at the conference saw little cause for worry.

An exception is Watson's (1971) exemplary concern to bring some of the problems associated with the cloning of humans to the attention of the general public. But it is difficult to take seriously his solution of avoiding problems by legislation and international agreement. Legislation (in tax matters, for example) clearly favors the dominant sectors over the weaker ones, and the record of Great Powers in international agreements hardly inspires confidence—in the case of America, the 1954 Geneva Accords on Indochina; or, more to the point, the 1899 Hague International Peace Conference banning the use of "asphyxiating or deleterious gases" and the 1925 Geneva Protocol on chemical and biological warfare, both of which clearly outlaw the kinds of chemical war we have waged in Indochina.

I have tried to show that *the structure of technological society in America, and particularly the role of science, are such as almost to preclude avoiding the monopolization of scientific knowledge by the dominant sectors for exploitation and war.* It is difficult to see how this situation will be changed by anything short of the most radical social transformation. If responsibility for the application of science should be borne not by science but by society as a whole, then responsibility will be assumed only when society as a whole is transformed. The dynamic of expanding technology without regard to consequence will have to be stopped; the power of the tiny technocratic minority will have to be redistributed; and the values defining how science is used will have to be changed. Only then can science be applied for the benefit and in the interests of all sectors of society, not primarily a ruling elite, and—most important—according to the thoughtful and responsible assessment of needs, not by a higher authority, but by those concerned.

Power is seldom yielded willingly, and the struggle will be a bitter one. But an analysis of science is hardly the only road to these conclusions, and more and more people, particularly the young and disadvantaged, who have the smallest stake in the present arrangement, are finding their way there.

A natural response is, what can we do as scientists? In a way, this already begs the question. Technocracy depends in some measure on separating the scientist class from society as a whole, and persuading these priests of the high culture that their interests lie with the ruling elite rather than ordinary people. This notion will have to be the first to go if there is to be any real progress.

In this context, two of the most common responses of scientists to questions of social responsibility are not only useless but counterproductive. One is the attempt to influence government from within, usually by personal contact. Technocracy is not likely suddenly to turn benevolent and relinquish control, and the titillations of the corridors of power invariably destroy perspective. In any case, negotiations are the province of society as a whole, not of the privileged scientific elite.

The other response is through the congressional lobbies and public relations channels that have proliferated since dissent has become acceptable. An example is the recently revitalized Federation of American Scientists—"no other group," says MIT President Wiesner (1970), "so truly represents the conscience of the American scientist," praise from a dubious source in view of the weapons and counterinsurgency research characterizing MIT and Wiesner's own career as presidential science adviser. By acting as a loyal rather than a critical opposition, such groups accept, and are often apologists for, technocracy's power and its definition of the role of science, and help restrict dissent critical of these fundamentals. In the recent controversy over the ABM, groups like this accepted without question the ground rules limiting disagreement to purely technical considerations of firepower, performance, relative military capability, and other expertise, thereby removing the problem from the purview of the ordinary person, at least as affected as the experts by the solution. This approach effectively headed off any critical appraisal of the real genesis of the ABM problem, namely the assumptions underlying the Cold War. Most scientists' groups opposing chemical and biological warfare or the use of herbicides in Vietnam have also taken this approach, and attempts to control gene manipulation will probably continue in the same conventional, nonperturbing terms.

Properly, then, the target is society as a whole, and the thrust should be mainly in the larger social arena, to break the power of the technocracy. But this is not to grant scientists leave to carry on business as usual. Like any other sector, scientists must begin developing an enlightened definition of the role of their work as part of a technological society free of technocracy, in which all sectors participate equitably. This will require a radical restructuring both of values and, given technology as an institutional system, of institutions, in science no less than elsewhere. Provided scientists recognize and combat the elitism built into the profession as it now stands, certain steps can be taken, not as patchwork reform, but as the beginning of a genuine solution (see also Beckwith, 1970; Zimmerman *et al.*, 1971).

1. *Scientific values.* Scientists should begin to investigate the character, role and implicit ideology of both their own work and science in general, rather than accepting technocracy's definition, rewarding though it may be. Biological scientists might begin analyzing how advances in such areas as medical care and agriculture are actually used, and exposing the ideology of pseudoscientific arguments purporting, say, to show racial differences in intelligence, which, quite aside from their intellectual weakness, intentionally obscure a rational consideration of the treatment of exploited sectors. Every university course should include a consideration of the social effects of the technical subject matter. For gene manipulation, the role of biological research in medicine and in turn of medical care in society should be thoroughly investigated, and those working on gene manipulation should be forced to confront the likely applications of their research. How interest, importance, and intellectual challenge are defined for the scientific community should be brought to light. And for the scientist, self-fulfillment as an individual will have to give way to fulfillment as a responsible member of society. All this will be difficult, since technocracy will respond with the accusation of anti-intellectualism.

2. *Scientific research.* Scientists must be aware, insofar as possible even before starting it, of the potential social consequences of research, especially in areas such as gene manipulations. Ultimately this will require recognizing that science can be used in conflicting ways in the interests of different sectors of society, and that scientists, rather than participating only in a disinterested search for truth, are by the research they choose to do implicitly advocating the interests of one sector or another. This will in turn shift the focus of research toward the needs of under-

privileged sectors here and exploited ones abroad, as defined not by the scientists but by those sectors themselves. Scientists could directly provide health-related research and training to ghetto groups, for example, or agricultural techniques to peasants (rather than despotic bureaucracies) in underdeveloped countries. A moral imperative supports the need in Indochina for research to help offset the effects of technology already applied there, in medical care for pellet bomb victims, reforestation of defoliated areas, rehabilitation of soils contaminated with herbicides, and so on. Applied science, then, will have to be developed and acknowledged as legitimate by scientists who now overvalue the prestige and challenge of basic research, if the weaker sectors are to have "their" science rather than being manipulated by the dominant sectors' science. These suggestions will also be difficult for scientists, who will have to abandon their accustomed professional rewards and learn to depend on new ones.

3. *Relations with other sectors*, probably most important of all. First of all, scientists can begin demystifying science for the nonscientist, not in the patronizing, top-down way science is now popularized, but rather by providing information in understandable form to those who need it. Poor women should be given information on reproductive physiology, say, or factory workers on industrial hazards, in clear language without the jargon and mystique of expertise. For gene manipulation this means openly and honestly presenting all the possibilities to the general public, and realistically accepting fears of "impersonal scientific manipulation." But further, it means searching for a solution, not among scientists or technocrats, but in a dialogue with the community at large, ultimately as part of a general dialogue on the role of science. Scientists can initiate the dialogue by clarifying technical information and soliciting the participation of the weaker sectors, who until now have had no say at all—and will probably be justifiably sceptical of the initial attempts. But scientists will have to abandon their intellectual arrogance and their self-image as an elite privy to knowledge ordinary people cannot possibly understand. This will be most difficult of all, since the powerful sectors will perceive this—and correctly so—as a serious threat to the elitism and privilege that help maintain them.

These are only beginnings. If they seem limited it is partly because of the scarcity of people thinking seriously about the problem. They amount to a basic acknowledgement of the adversary relationship between the technocratic minority which controls resources and the weaker sectors who have little control over their own lives, and of the role of science in maintaining it. Unless there is radical change in the context in which science is done and its benefits applied, science will continue to serve only the minority, and there will be little hope of controlling gene manipulation or any other scientific advance.

ACKNOWLEDGMENT

I thank Fred Ausubel, Noam Chomsky, Ira Herskowitz, Frank Mirer, Phil Myers, Barbara Signer, and Bill Zimmerman for criticizing the manuscript; Bill Zimmerman and Len Radinsky for showing me their paper before publication; Pham Van Bach for Fig. 3, and Salome Waelsch for the remaining figures; and the National Institutes of Health and the American Cancer Society for support.

REFERENCES

AGARWAL, K. L., BÜCHI, H., CARUTHERS, M. H., GUPTA, N., KHORANA, H. G., KLEPPE, K., KUMAR, A., OHTSUKA, E., RAJBHANDARY, U. L., VAN DE SANDE, J. H., SGARAMELLA, V., WEBER, H., and YAMADA, T. (1970). Total synthesis of the gene for an analine transfer ribonucleic acid. *Nature (London)* **227**, 27–34.

ALLISON, A. C. (1954). Protection afforded by sickle-cell trait against subtertian malarial infection. *Brit. Med. J.* **1**, 290–294.

APOSHIAN, H. V. (1971). The use of DNA for gene therapy—the need, experimental approach, and implications. *Perspect. Biol. Med.* **14.**

ASIMOV, I. (1971). The scientists' responsibility. *Chem. & Eng. News* **52**, 3–7.

BALTIMORE, D. (1970). An RNA-dependent DNA polymerase in virions of RNA tumor viruses. *Nature (London)* **226**, 1209–1211.

BARAN, P. A., and SWEEZY, P. M. (1966). "Monopoly Capital." Monthly Review Press, New York.

BECKWITH, J. (1970). Gene expression in bacteria and some concerns about the misuse of science. *Bacteriol. Rev.* **34**, 222–227.

BECKWITH, J. (1971). The social impact of modern biology. *Science for the People* **3**, No. 2, 1–8.

BECKWITH, J., and ZIPSER, D., eds. (1970). "The Lactose Operon." Cold Spring Harbor, New York.

BECKWITH, J. R., SIGNER, E. R., and EPSTEIN, W. (1966). Transposition of the *lac* region of *E. coli. Cold Spring Harbor Symp. Quant. Biol.* **31**, 393–401.

BRUNNER, J. (1969). "Stand on Zanzibar." Ballantine, New York.

COURTNEY, K. D., GAYLOR, D. W., HOGAN, M. D., FALK, H. L., BATES, R. R., and MITCHELL, I. (1970). Teratogenic evaluation of 2,4,5-T. *Science* **168**, 864–866.

Daedalus. (1969). Ethical aspects of experimentation with human subjects. *Daedalus (Boston)* **98**, No. 2.

DAVIS, B. D. (1970). Prospects for genetic intervention in man. *Science* **170**, 1279–1283.

DOMHOFF, G. W. (1967). "Who Rules America?" Prentice-Hall, Englewood Cliffs, New Jersey.

EDWARDS, R. G., STEPTOE, P. C., and PURDY, J. M. (1970). Fertilization and cleavage *in vitro* of preovulatory human oocytes. *Nature (London)* **227** 1307–1309.

EHRENREICH, B., and EHRENREICH, J. (1971). "The American Health Empire: Power, Profits and Politics." Random House, New York.

FLEMING, D. (1969). On living in a biological revolution. *Atl. Mon.* Jan., 64–70.

FLEMING, D. (1970). Big science under fire. *Atlantic* Sept., 96–101.

GALSTON, A. W. (1971). Education of a scientific innocent. *Natur. Hist., N.Y.* **80,** No. 6, 16–22.

GARDNER, R. L., and EDWARDS, R. B. (1968). Control of the sex ratio at full term in the rabbit by transferring sexed blastocysts. *Nature (London)* **218,** 346–348.

GLASSMAN, J. K. (1970). Harvard genetics researcher quits science for politics. *Science* **167,** 963–964.

GOTTESMAN, S., and BECKWITH, J. R. (1969). Directed transposition of the arabinose operon: A technique for the isolation of specialized transducing bacteriophages for any *Escherichia coli* gene. *J. Mol. Biol.* **44,** 117.

GREENBERG, D. S. (1967). "The Politics of Pure Science." New American Library, New York.

GURDON, J. B., and UEHLINGER, V. (1966). "Fertile" intestinal nuclei. *Nature (London)* **210,** 1240–1241.

HAYES, W. (1968). "Genetics of Bacteria and their Viruses," 2nd ed. Wiley, New York.

HILL, M., and HILLOVA, J. (1971). Recombinational events between exogenous mouse DNA and newly synthesized DNA strands of chicken cells in culture. *Nature (London), New Biol.* **231,** 261–265.

HOROWITZ, D. (1965). "The Free World Colossus." Hill & Wang, New York.

JONES, R. C. (1971). Uses of artificial insemination. *Nature (London)* **229,** 534–537.

LEDERBERG, J. (1970). Genetic engineering and amelioration of genetic defect. *Bio-Science* **20,** 1307–1310.

LEDOUX, L., and HUART, R. (1969). Fate of exogenous bacterial deoxyribonucleic acids in barley seedlings. *J. Mol. Biol.* **43,** 243–262.

LEWONTIN, R. (1970). University research and the professional class. *NUC Pap.* **2,** 3–7.

LUNDBERG, F. (1968). "The Rich and the Super-Rich." Lyle Stuart, New York.

LURIA, S. E. (1972). Science, technology and responsibility. *Proc. Amer. Phil. Soc.* **116,** 351–356.

MCDERMOTT, J. (1969). Technology: The opiate of the intellectuals. *N.Y. Rev.* **13,** 25–35.

MAGDOFF, H. (1969). "The Age of Imperialism." Monthly Review Press, New York.

MICHAELSON, M. (1971). The coming medical war. *N.Y. Rev.* **16,** 32–38.

MILLS, C. W. (1956). "The Power Elite." Oxford Univ. Press, London and New York.

MILUNSKY, A., LITTLEFIELD, J. W., KANFER, J. N., KOLODNY, E. H., SHIH, V. E., and ATKINS, L. (1970). Prenatal diagnosis. *N. Engl. J. Med.* **283,** 1370–1381, 1441–1447, and 1498–1504.

MINTZ, B. (1967). Gene control of mammalian pigmentary differentiation. I. Clonal origin of melanocytes. *Proc. Nat. Acad. Sci. U.S.* **58,** 344–351.

Nature. (1970a). Editorials and letters. *Nature (London)* **224,** 834–835, 1241–1242, and 1337.

Nature. (1970b). Editorials and letters. *Nature (London)* **225,** 301–302.

Nature. (1970c). The biologist's dilemmas. *Nature (London)* **228,** 900–901.

Nature. (1971). PSAC hiccoughs over 2,4,5-T. *Nature (London)* **231,** 210–211.

Nature, New Biology. (1971a). Can genetic defects be corrected in cells? *Nature (London), New Biol.* **230,** 1–2.

Nature, New Biology. (1971b). First steps to genetic engineering? *Nature (London), New Biol.* **231,** 257–258.

OSTERMAN, N., WADDELL, A., and APOSHIAN, H. V. (1970). DNA and gene therapy: Uncoating of polyoma pseudovirus in mouse embryo cells. *Proc. Nat. Acad. Sci. U.S.* **67**, 37–40.

PLATT, J. (1969). What we must do. *Science* **166**, 115.

ROGERS, S. (1970). Skills for genetic engineers. *New Sci.* Jan., 195–196.

ROSE, M. (1970). Pangloss and Jeremiah in science. *Nature (London)* **229**, 459–462.

SAMBROOK, J., WESTPHAL, H., SRINIVASAN, P. R., and DULBECCO, R. (1968). The integrated state of viral DNA in SV40-transformed cells. *Proc. Nat. Acad. Sci. U.S.* **60**, 1288–1295.

SCHWARTZ, A. G., COOK, P. R., and HARRIS, H. (1971). Correction of a genetic defect in a mammalian cell. *Nature (London) New Biol.* **230**, 5–8.

Science. (1970a). Letters. *Science* **167**, 1668–1669.

Science. (1970b). Letters. *Science* **168**, 1285.

Science. (1971). Herbicides in Vietnam: AAAS study finds widespread devastation. *Science* **171**, 43–47.

SHAPIRO, J., MACHATTIE, L., ERON, L., IHLER, G., IPPEN, K., BECKWITH, J., ARDITTI R., REZNIKOFF, W., and MACGILLIVRAY, R. (1969). The isolation of pure *lac* operon DNA. *Nature (London)* **224**, 768.

SIEKEVITZ, P. (1970). Scientific responsibility. *Nature (London)* **227**, 1301–1303.

SINHEIMER, R. L. (1969). The prospects of designed genetic change. *Engin. Sci. Magazine, Cal. Inst. Techn., April 1969,* 15–20.

SZYBALSKA, E. H., and SZYBALSKI, W. (1962). Genetics of human cell lines. IV. DNA-mediated heritable transformation of a biochemical trait. *Proc. Nat. Acad. Sci. U.S.* **48**, 2026–2034.

TELLER, E. (1971). Cited in *Science for the People* **3**, No. 1, 10.

TEMIN, H. M., and MIZUTANI, S. (1970). RNA-dependent DNA polymerase in virions of Rous Sarcoma virus. *Nature (London)* **226**, 1211–1213.

Time Magazine. (1971). Man into superman. The promise and peril of the new genetics. *Time Mag.* April, 33–52.

TITMUSS, R. (1971). "The Gift Relationship." Pantheon Books, Random House, New York.

TOMKINS, G. M., GELEHRTER, T. D., GRANNER, D., MARTIN, D., SAMUELS, H. H., and THOMSON, E. B. (1969). Control of specific gene expression in higher organisms. *Science* **166**, 1474–1480.

TUNG, T. T., AUH, T. K., TUYEN, B. Q., TRA, D. Z., and HUYEN, N. Z. (1970). Les effects cliniques de l'utilisation massive et continue de defoliants sur la population civile. Documents de la Réunion Internationale de Scientifiques sur la Guerre Chimique au Vietnam, Orsay (Paris), Décembre, 1970. *In* "Guerre Chimique," Etudes Vietnamiennes No. 29, pp. 12–14.

WATSON, J. D. (1968). "The Double Helix." Athenaeum, New York.

WATSON, J. D. (1971). Moving toward the clonal man: An example of scientific inevitability? *Atl. Mon.* **227**, 52–53.

WEINBERG, A. M. (1970). In defense of science. *Science* **167**, 141.

WHITTINGHAM, D. G. (1968). Fertilization of mouse eggs *in vitro*. *Nature (London* **220**, 592–593.

WIESNER, J. (1970). "Organizing Brochure." Federation of American Scientists.

WILSON, M. (1970). On being a scientist. *Atlantic* Sept., 101–106.

ZIMMERMAN, B., RADINSKY, L., ROTHENBERG, M., and MEYERS, B. (1972). "Towards a Science for the People." *Science for the People.* Jamaica Plain, Massachusetts.

Subject Index

A 4
B 5
C 6
D 7
E 8
F 9
G 0
H 1
I 2
J 3